Praise for Leading for Organisational Change

Working to support senior executives and boards on a daily basis, the most common themes we see are how they grapple with, or sometimes resist the challenges of change, and critically how they can find a way to best lead their people. This inspiring book shines a light on a way forward and crucially, it is brilliantly practical. It will undoubtedly captivate an executive and provide real pause for reflection.

—Mark Franklin, CEO and accredited business coach of Freefort

Emery delivers a clear-eyed, behind-the-scenes account of a critical juncture in a firm's development. She uses it to draw out practical insights that should resonate with every business leader seeking to inspire organisational change.

—Dr. Heidi K. Gardner, Distinguished Fellow
at Harvard Law School

A shining light where business books are either drab or over-simplistic, Jennifer Emery speaks to the heart of what modern business should be - purposeful and human.

Her approach is intelligent, multi-layered and people-orientated, much like the change management approach she proposes. At the heart of her approach is the desire to weave people together through storytelling, which she models through the surprisingly inspiring story of the ambitious merger of law giant CMS.

A delight to read - and a very useful delight at that.

—Sarah Lloyd-Hughes
Leadership Communications Coach, speaker & author,
"How to be Brilliant at Public Speaking"

Written by a lawyer, you'd expect this book to be impeccably researched, evidence-based, and brilliantly argued. What you might not expect is a story with so much heart. Emery's examination of leading organisational change, both rigorous and romantic, will show you how to effect change and create the desire to make it happen.

—Richard Hytner, Adjunct Professor of Marketing,
London Business School and founder of beta baboon

LEADING FOR
ORGANISATIONAL CHANGE

LEADING FOR ORGANISATIONAL CHANGE

BUILDING PURPOSE, MOTIVATION AND BELONGING

JENNIFER EMERY

WILEY

This edition first published 2019
© 2019 Jennifer Emery

Registered office
John Wiley & Sons Ltd, The Atrium, Southern Gate, Chichester, West Sussex, PO19 8SQ,
United Kingdom

For details of our global editorial offices, for customer services and for information about
how to apply for permission to reuse the copyright material in this book please see our
website at www.wiley.com.

Wiley publishes in a variety of print and electronic formats and by print-on-demand. Some
material included with standard print versions of this book may not be included in e-books or
in print-on-demand. If this book refers to media such as a CD or DVD that is not included in the
version you purchased, you may download this material at http://booksupport.wiley.com. For
more information about Wiley products, visit www.wiley.com.

Designations used by companies to distinguish their products are often claimed as trademarks.
All brand names and product names used in this book are trade names, service marks,
trademarks or registered trademarks of their respective owners. The publisher is not associated
with any product or vendor mentioned in this book.

Limit of Liability/Disclaimer of Warranty: While the publisher and author have used their best
efforts in preparing this book, they make no representations or warranties with respect to the
accuracy or completeness of the contents of this book and specifically disclaim any implied
warranties of merchantability or fitness for a particular purpose. It is sold on the understanding
that the publisher is not engaged in rendering professional services and neither the publisher
nor the author shall be liable for damages arising herefrom. If professional advice or other
expert assistance is required, the services of a competent professional should be sought.

Library of Congress Cataloging-in-Publication Data

Names: Emery, Jennifer, 1977- author.
Title: Leading for organisational change : building purpose, motivation and
    belonging / Jennifer Emery.
Description: First Edition. | Hoboken : Wiley, 2019. | Includes
    bibliographical references and index. |
Identifiers: LCCN 2018057472 (print) | LCCN 2018060127 (ebook) | ISBN
    9781119517986 (ePub) | ISBN 9781119517955 (ePDF) | ISBN 9781119517962
    (hardback)
Subjects: LCSH: Organizational change. | Employee motivation. | Personnel
    management. | BISAC: MATHEMATICS / Applied.
Classification: LCC HD58.8 (ebook) | LCC HD58.8 .E4624 2019 (print) | DDC
    658.4/06—dc23
LC record available at https://lccn.loc.gov/2018057472

Cover Design: Wiley
Cover Image: © Jakub Grygier/Shutterstock

Set in 10/14.5pt Palatino by SPi Global, Chennai, India
Printed in Great Britain by TJ International Ltd, Padstow, Cornwall, UK

10 9 8 7 6 5 4 3 2 1

To Tommy, Frankie, Esther and Gabriel – my purpose.

If you want to build a ship, don't drum up the people to gather wood, divide the work and give orders. Instead, teach them to yearn for the vast and endless sea.

Antoine de Saint-Exupéry

CONTENTS

INTRODUCTION

I dwell in possibility

Emily Dickinson

Why You Need This Book

When I was fourteen, I got my first Saturday job in a little shop in Edinburgh that sold furnishing fabric. This job taught me how to match a coloured chintz lining to a floral, how to count in 64cm pattern repeats, and how to make a mean bacon roll. It gave me biceps to die for from lugging around big bales of fabric. It also taught me that people do not like change. The shop was owned by a brilliant female entrepreneur. She opened one shop, she opened another. She moved to bigger premises. She branched out from dress fabric into curtain fabric. She experimented with telephone ordering (there was no internet then). Every single time she tried anything new, one of the other women who worked there, or a customer, would suck their teeth, say how much they'd preferred things the old way and darkly predict the imminent demise of the business. That was in 1991. The business is still going strong.

Since then, throughout my entire career, I have been around businesses that are about to change, are changing, or have just changed. And yet the whole rhetoric around change implies that it is an aberration. The very language we use pathologises change and implies it is something to be feared, dodged, and if not dodged then minimised and managed. If this is true, then it is bad news indeed, because we read everywhere that we live in VUCA times – that is to say, the world is volatile, uncertain, complex and ambiguous. Change, we are told, is the new constant.

1

This book presents a bold alternative view of change. It argues that, done well, change can be a positive thing. That it is not change *per se* that presents a problem for organisations and destroys value, but rather some common underlying assumptions about what change means, and a resulting tendency on the part of those leading change projects to put undue emphasis on some things, and to skate over others. That it is possible for organisations and the individuals within them not just to survive periods of change but actively to thrive throughout, and to emerge stronger as a result.

I am a person who loves to read and learn. The sort of person who, when faced with any new situation, would prefer to sink £100 on Amazon before taking a single practical step. Every time I have been faced with a new challenge around change – a reorganisation, a rebrand, or, perhaps most notably, a merger or acquisition or disposal of some kind – I have hungrily foraged for all the wisdom that is out there in books, journals, podcasts and TED talks. Over the years, as my own experience has widened and my understanding of what 'good' change looks like has deepened, I have become increasingly intrigued and discouraged by the paucity and the dryness of the material out there on the challenges and opportunities that change can bring. There is plenty about the technical and financial aspects of managing mergers or reorganisations, plenty about systems and processes and project management, but nothing which really speaks to the messy human reality of what organisations are grappling with when leading hundreds, perhaps thousands, of human beings through a period of profound change. I am always deeply grateful when I do encounter something relevant and engaging, but I have found I have had to dig hard for it – drawing from psychology, neurology, sociology, and then extrapolating from there. It is a struggle and not one I have had endless time for, and so my reading over the years, though wide, has tended towards the utilitarian and somewhat superficial.[1] I have long been itching both to re-explore some of the terrain I have been dashing through, and to fill the gap I have perceived in the literature.

In the end, though, I was ultimately inspired to write this book by virtue of having had the opportunity to participate in a period of profound organisational change that more closely exemplified best practice in this area than anything I have seen or been involved with previously. In early 2016, I was asked to get involved in the team that would put together and then implement from a people and cultural perspective what became the three-way merger of CMS, Nabarro and Olswang to create a new top five global law firm.

Now. Report after report will tell you, almost irrespective of when you are reading this, that right now, no, right NOW, we have hit an all-time high for M&A activity across the globe.[2] And yet study after study will also tell you that the failure rate for mergers is anywhere between 70% and 90%.[3] Clearly, something is going drastically wrong. M&A is one of the riskiest investments for organisations of their time, money and reputations – and yet organisations worldwide press ahead with this strategy for growth, convinced that they will be the ones to buck the trend.

Commentators are increasingly identifying an underestimation of, and a lack of investment in, the people and cultural aspects of an M&A process as one of the main reasons for failure.

It is those people and cultural aspects that this book is all about.

Almost my entire career has been spent in and around professional services firms – first as a corporate lawyer, and then as a coach, strategist and leader. Professional services organisations are generally 'expert' cultures – full of highly intelligent people with a deep interest in, and commitment to, their particular vocation or field of expertise – lawyers, accountants, surveyors, architects, engineers – people for whom deep specialism is critical. Professional services organisations also tend to be reasonably flat in terms of hierarchy and to operate primarily by consensus and collaboration rather than command and control. This is often also reflected in their ownership structures – such organisations are often structured as partnerships, or shared ownership models. It is perhaps so obvious as to go without saying, but the other striking feature of professional services organisations is that their 'product' is their people. This means that professional services firms are only as good as their capacity to attract talent and then engage and equip that talent to work, individually and collectively, to solve the problems, create the products and deliver the services that the market wants to buy from them.

I have come to understand that, precisely because of these particular features, experience of change gained in the particular context of professional services is particularly rich in learning. If people are your product, if your entire business model depends on their being engaged, resourceful and creative, then it is surely vital to ensure that any change process works in such a way as to, at the very least, create minimal adverse impact for them, and ideally in such a way as to actively enable them to flourish and grow. If you are operating within a flat hierarchy, with highly intelligent people, many of whom are also owners of the business, then an inclusive, nuanced and iterative approach to change is an imperative.

There is also a point about scale. Global law firms in particular, while enormous in terms of their reach and influence, are smaller in headcount terms than the global clients they serve. This makes it possible for a team involved in a change process in a law firm to both wrestle with complicated details and still zoom out far enough to get their arms around the entire big picture. This brings a unique perspective on the human aspects of merger.

All of this meant that the chance to get involved in the CMS-Nabarro-Olswang merger felt like the opportunity of a lifetime in change terms. Addressing the people and cultural issues – precisely the issues which are so consistently and gravely overlooked in a larger corporate context and so poorly addressed in the current literature – was an absolute imperative in this case, and the leadership teams across all three firms understood that intimately right from the outset.

The opportunity of a lifetime it proved to be. The new firm created by the merger is a product almost entirely of its own creation, put together under the strategic leadership and guidance of its senior management team, and integrated operationally by its own people. The leadership team understood that it was important that this process itself should become part of the story of the new firm, and that it would be by virtue of going through the process together that people would best be able to understand one another's existing stories, cultures and values and weave them into something new. The team delivered a co-located, single firm, operating as one team, on one set of systems, and delivering a fully integrated service to clients right from Day 1 – an unprecedented and astonishing achievement. In its first year, the new firm achieved a 19% increase in profits, exceptional client feedback and extensive industry recognition, flying in the face of the M&A trends. The central focus and emphasis that the leadership placed on people and culture was one of the most – perhaps *the* most – critical factors in the new firm's success.

The CMS merger epitomises much of what I have come to believe is the very best approach to leading organisational change. For this reason, as well as being the final inspiring nudge for me to write this book, it has also become the featured 'case study' or exemplar, and I will return to the story of the CMS merger and the learning and inspiration to be drawn from it throughout the book. I do so with the support of the CMS leadership team, but the observations and learnings that follow are my own. This is not 'the official story' of the CMS merger; that is for others, or history, to write.

The central thesis of this book, in a nutshell, is that an absolute focus on people and culture, complete clarity on purpose and an approach rooted in storytelling can enable businesses to navigate periods of profound change, including merger, in such a way as to actively generate value and enable organisations and individuals to flourish. There is a quiet revolution in this message. The widely accepted, and entirely understandable, response in times of uncertainty and change is to take control – to impose more structure, more systems, more rules, policies and reporting, to minimise risk and ambiguity and close down doubt. This book advocates for a different approach, one which is subtler and requires patience, and which focuses on changing behaviour by changing minds and hearts. It requires letting go when you want to hold on; opening up when you want to close down. It allows space for ambiguity and for mistakes. It is about planting seeds and letting them grow.

This feels to me an important and timely thesis. I believe there to be a profound and fascinating societal shift underway which makes it important for organisations to begin broadening their perspective, putting their people and culture at the heart of everything, and grappling with questions around what value really means, the nature of their underlying purpose, and the basis upon which they engage with their stakeholders. As the patriarchal power structures of the industrial age are buckling, approaches to leadership and management that depend on command and control and 'holding on' are being replaced by approaches which depend upon empowering people and building consensus. This book begins to explore some of that and hopefully adds to the 'business book' canon a more pragmatic, human and holistic voice than is perhaps the norm.

Not enough is written about our working lives generally. I guess I am thinking about office work in particular, but I think the same is true across the board.[4] But perhaps that is because we don't even talk with one another much about our work. For something that takes up so much of our lives, the places we go each day, the things we fill our time with, the people we sit alongside – are oddly missing from the stories we tell. Distilled down to trite little 'Dilbert' narratives, or drinks down the pub at Christmas. Whether because we are oddly coy about it, or possessive, or cynical, or beset by the suspicion that this, of all experiences, is not one that is universal and capable of being shared or understood by anyone who is not there, we downplay our work lives, overlooking or ignoring the fact that this, too, is part of our story. Here, too, is the ground we stand on, the difference we make. This is where we win and lose, help and hinder, learn and teach. Where we try new things, sometimes fail, witness close up the lives, loves and losses of other people who we are in community with. This is where we live in sight and sound of each other. This is where we are church and community. This is where we do life.

We are living in times where constant change is the new normal. If we want our organisations to thrive, if we want to build systems and structures that enable us to do business, to exchange ideas and goods, to create new things, to tackle challenges and hardships and inequalities, then we need a people-centred approach to our organisations. We need to focus on our purpose, our culture and values, and the stories we tell.

How to Use This Book

This book is in four parts.

Part One of this book introduces some ideas and theories. There are some thoughts around the role of storytelling, and quite a lot of theory around the definition and importance of purpose, culture and values. Then there is some theory around how change happens, individually and corporately, and finally there is some neuroscience and behavioural science about what is actually happening to us personally when change is happening around us.

Then, in Part Two, I share a little of the story of the CMS merger. This serves three purposes – it provides a bridge between the theory in Part One and the practical application in Part Three, giving context and background; it sets out in one place those parts of the CMS merger experience which best illustrate the approach to change in that case, and its impact; and it exemplifies the power of storytelling – a central tenet of the book. If you are all over the theory, not interested, or plan to loop back later, you could just dive straight in here. This second part of the book is devoted to simply telling a story – again, this is *my* story or version of events, not *the* official CMS story. I hope it is interesting enough that you will forgive its vanities and elisions. This is the story of one aspect in particular (people and culture) of one merger in particular, told by one person in particular, and so it is as subjective, biased and full of holes as you might expect. So are all the other stories that you will find in this book. They are eyewitness accounts, anecdotes and survivors' tales. Individually and collectively, these stories contain experiences and ideas that can inspire and instruct.

Part Three gets practical. It draws out some of the key learnings and themes around the people aspects of organisational change, drawing on both the CMS merger story and some different stories from different circumstances in different industries. This part explores how, in practice, an approach to change which is centred on purpose and values, and relies heavily on the creation and telling of stories, can help organisations to rise to some of the biggest challenges and opportunities that periods of profound change present. If you are wrestling with challenges and want a fast track to some ideas on how to tackle them, you might start with this third part and work backwards.

Finally, Part Four of the book contains some reflections and ideas about the context within which the ideas posited in this book exist – the wider systemic and societal factors. It considers the implications for leadership if the ideas contained in this book are to be sustainable beyond the particular context in which they are presented. What do leaders need to do and be in order to foster this purpose-driven, story-rich approach in their organisations and teams?

This is not a 'how to' book in the sense that it does not have checklists and tick-box exercises. The whole point of the book, really, is that how one organisation manages change will be unique to that organisation, and not susceptible of being distilled into a list and re-used, parrot fashion, elsewhere. We can also, I think, take technical competence in this area as a given. The people responsible for making change happen in organisations generally know how to manage a project, transact a deal, allocate and manage risks, plan logistics, etc. This book is intended to speak into the gap I perceived in the literature – namely, how to grapple with the human aspects and to better equip everyone in organisations to be able to thrive during periods of change.

So, not a 'how to' book – but hopefully a 'helpful' book – blending the best theory from a broad range of sources, with deep and recent practical experience, and presenting a framework within which individuals can think through change in their own organisations. This book will be helpful if you are involved in transforming businesses, and are interested in making the new business a success. This book will resonate if you are alive to the impact of change and disruption on the wellbeing and performance of individuals and teams. You will be interested in this book if, more generally, you are casting around for a more purposeful and emergent approach to leadership, and if you are interested in all the talk around 'purpose' but want to understand more about how it drives value.

It is also intended to be a deeply honest book – I wanted to show the learning in real-time and to think aloud and in an accessible way about how the theory works in practice. Many of the books I love best in this field have been written by academics – by people whose job it is to think clever thoughts, do research to test those thoughts, read other people's clever thoughts and then commit the whole lot to paper. Tucked away in the acknowledgements section at the back of these books, there is often a sentence which says something like, 'This book has been ten years in the making.'

I am so grateful for, and inspired by, the profound work and thinking in these books. But this approach is unfeasible for practitioners. It perhaps goes without saying that the book you are now holding in your hands was *not* ten years in the making. If we want to hear practitioners' voices tell us about real, lived experiences from the thick of the action, we have to be relaxed about that. This is one perspective, honestly delivered, while the memories are still fresh, and the lessons still real.

This book was around six months in the making, squeezed around a full-time, full-on senior role and four small children. It has been written in long and short snatches – in early mornings with coffee and late nights with wine, in airport lounges, on packed commuter trains, outside ballet classes and on the touchline during an under-12s football tour in Denmark in blistering heat.

I listened recently to a podcast in which Professor Adam Grant, a leading organisational psychologist, was discussing the huge rise in popularity of accessible writing on social science, which perhaps started with the widespread interest in Malcolm Gladwell's work. Grant observed that this has led to an increase in the number of compelling and accessible books in the fields of sociology, social psychology, behavioural economics, and speculated that perhaps soon, finally, an evidence-based, academically rigorous approach would also come to prevail in the popular writing on organisational dynamics, a field which is currently dominated by practitioners speaking from the echo chamber of their own experience. Ah well, I'm either the last of a dying breed or a bridge between old and new. As G.K. Chesterton reputedly said, 'If a thing is worth doing, it's worth doing badly.'

If I do not, in the end, believe this to be a book 'done badly', that is because it is standing on the shoulders of giants. It is founded in some rigorous reading and for the most part tries to substantiate any assertions it makes with some sort of reference or evidence. It is also founded in two decades of personal experience and on-the-job learning from the countless incredible leaders, mentors and peers whom I have been blessed work alongside.

One final observation. I have struggled throughout the writing of this book with a bad case of imposter syndrome. I found as I embarked on writing that I did not want to 'assert', I wanted to discuss. This meant, though, that early drafts read like a first-year university essay.

I studied the pile of books around my desk to see how others have tackled this. Some people just assert right from the off. You can be eighty pages deep in some astonishing new theory before even thinking to check whether any of it is remotely based in experience or research. Others are closer to the discursive essay end of the spectrum. The vast majority of the latter group are women. This is interesting.[5]

In the end, I have tried to strike a middle ground. Where I have an opinion, I want to share it. Unapologetically. I will occasionally 'assert'. But I also want to bring you the nuance and breadth of the thinking that is out there and leave space for you to reflect and draw your own conclusions.

Writing this book has brought me great joy and rich learnings of its own. I offer it lightly, and in the hope that it adds to the sum of knowledge.

Jennifer Emery,
London, July 2018

Notes

1. The bibliography for this book is full of rich resources and all have something to offer, but for my money my 'Hall of Fame' of books that I have gleaned most from and relied most heavily about reads as follows:
   Fast/Forward by Julian Birkinshaw and Jonas Ridderstrale
   Positive Professionals by Anne Brafford

> *Daring Greatly* by Brené Brown
> *Alive at Work* by Dan Cable
> *The Sweet Spot* by Christine Carter
> *Be More Pirate* by Sam Coniff Allende
> *Beyond Measure* by Margaret Heffernan
> *Reinventing Organisations* by Frederic Laloux
> *The Story Factor* by Annette Simmons
> *How to Have a Good Day* by Caroline Webb

2. ADP, *Achieving M&A Success*, at https://www.adp.co.uk/assets/vfs/Family-32/adp-files/Insights-Resources/Whitepapers/Docs/adp-unleashing-m-and-a-success-fy17.pdf

3. Christensen et al, The Big Idea: The New M&A Playbook, *Harvard Business Review*, March 2011

4. Though I have read *Intuition* by Allegra Goodman (Atlantic Books, 2010) about life in a laboratory. It's brilliant!

5. Because it is surely as important to be able to find role models in the world of business books as in the world of business itself, I have particularly tried to draw on and quote as many other women as I can – practitioners, academics and authors – in this course of this book. It has proven to be more difficult than it should be to find many female voices to choose from.

Part One

Foundation

Chapter 1

Once Upon A Time

Winter 2006. Early in the new year. The days short, grey and bitterly cold. And I was besotted. The first few months of my first son's life were a staggering love story. I couldn't get enough of his beautiful face, which changed every day and yet was constantly, uniquely and brilliantly him. Everything he did was fascinating. I would spend hours marvelling at all the potential condensed into his intense little body, and imagining a thousand bright futures for him. It was the most immense privilege to have played my part in his coming into the world, and I was both humbled and energised – I felt I could leap mountains for him.

The first few months of my first son's life were also a bracing roller coaster ride. Who was this demanding, noisy and irrational stranger? What had happened to my previously friendly body? Everything hurt or leaked. The house was a shambles and I was late for everything. None of the books talked about this. What was I doing wrong? There was also the teeth-grinding tedium of the routine. Eat, poo, sleep, repeat. My conversation shrivelled to repeating anecdotes from Radio 4's Women's Hour. My husband, family and friends were mysteriously replaced by irritating idiots.

Every day I would look at the clock convinced the day must be nearly over only to find it was barely lunchtime. I was sure a baby had felt like a good idea a year previously, but could not for the life of me remember why. I was very, very tired. I took a lot of baths and drank a lot of wine.

***

Why am I telling you this story? To make you smile. To paint a picture. To elicit empathy. To welcome you. To ignite your curiosity. To make you want to stay and read on.

Also, to set up a memorable metaphor . . .

Early summer 2017. The days warming, lengthening and loosening. And I was besotted. The first few months of our newly merged firm's life were a staggering love story . . .

There are lots of fun parallels between life in the eye of a major organisational change project and life in the eye of the ultimate life-change project that is the arrival of a baby, and we all love a good story. Staying up too late watching a box set (*just one more* . . .), reading by torchlight under the covers when our parents have switched the light off . . . we are echoing our ancestors around the campfire – listening, imagining, falling in love, slaying dragons, scaring ourselves silly, hanging off cliffs.

We are twenty-two times more likely to remember a story than a set of facts.

For a long time, psychologists thought that our proclivity for storytelling may be no more than what Steven Pinker[1] calls evolutionary 'cheesecake' – a fun, but ultimately useless, titillation for the restless machine that is our cognition. But a brilliant new study on hunter-gatherer societies proposes[2] that telling stories may in fact be an important mechanism by which knowledge is shared – the sort of 'who knows what about what' type of meta-knowledge that society needs us to have in order to function.

Stories are about the 'rules of the game' and the consequences of breaking them, just as gods with thunderbolts are for those of a more religious bent. And in much the same way, they help to promote cooperation and to encourage groups to bond. These findings chime with the theory of journalist and author Christopher Booker, who argues in *The Seven Basic Plots*[3] that we tell stories in order to pass a model for life down the generations.[4]

The central tenet of this book is that organisations need stories during periods of profound change. When things are volatile, or uncertain, or otherwise changing – *pretty much all the time, then!* – people need to make sense of the world around them. Groups of people need something to hold them together and help them move forward in a loosely coordinated way. Visions and strategy statements lose their power a little when volatility, uncertainty, conflict and ambiguity make it hard to discern the path ahead. Stories are less linear and can function at a different level. They can clarify and galvanise even when the times are uncertain and scary.

Not any old story will do, though. Not all stories are created equal. The best stories – the ones that give you goosebumps, or make you cry, or prompt you to go home and sell everything you have and pack a bag – those are the stories that speak straight to the heart of what it means to be us: who we are, why we're here, what matters, why we do what we do and how we do it. In other words, stories about our purpose and values.

So, the central tenet of this book, really, is that businesses need a purpose. And shared values. And stories – for the business as a whole, and for every business unit, team, office and human being within it – that speak into that purpose and those values and bring them to life; all shadows dancing around the same campfire of one single, bright common purpose. Those organisations – and their leaders – who are clear about why they are here and doing what they're doing . . . those businesses and leaders who can craft and tell stories that communicate those things well, individually and corporately . . . those are the businesses and leaders who are better able to drive long-term value by tackling the big challenges and grabbing hold of the big opportunities that periods of profound change present.

Defining Terms

In this book we are going to talk a lot about stories, culture, values and purpose. They are all connected ideas, but distinct. Each concept is unpacked in more detail a little later, but here, for now, is an attempt at a working definition of each of the various terms:

- **Purpose** – this is what an organisation is here to do. Why are these particular people, with these particular skills, experiences, relationships and assets, configured together into some sort of organisation? If an organisation vanished tomorrow, what would be missing in the world? This is purpose. Purpose is an organisation's North Star.

- **Values** – these are the guiding principles of an organisation. As the guys at Netflix say, 'Values are what we value.' They may be spoken or unspoken, but they are the non-negotiable things – the bright lines.

- **Stories** – these are a primary means by which purpose and values are articulated and explained, usually with some other things – strategy, successes, common experiences, artefacts – thrown in. They also, in turn inform purpose and values.

- **Culture** – this is the whole lot, and everything else. It is the air an organisation breathes; the soil it grows in. Culture is shaped by purpose and values, and by stories – but also by strategy, systems, history, the times we live in, where we are physically located, and the people leading us.

Notes

1. Pinker, Steven. *How the Mind Works*, Penguin, 1999.
2. Smith, Schlaepfer, Major, Dyble, Page, Thompson, Salali, Mace, Astete, Chaudhary, Ngales, Vinicius & Migliano. 'Cooperation and the Evolution of Hunter-Gatherer Storytelling' *Nature Communications* 8: 1853, 5 December 2017.
3. Booker, Christopher. *The Seven Basic Plots: Why we Tell Stories*, Continuum, 2005.
4. Wait, what? Seven basic plots? Yes. The central tenet of Booker's book is that the vast majority of all of our stories follow one of seven basic structures: overcoming a monster (Beowulf) . . . rags to riches (Cinderella) . . . the quest

(The Odyssey) . . . voyage and return (Watership Down) . . . comedy (Twelfth Night) . . . tragedy (Anna Karenina) . . . and rebirth (A Christmas Carol). It's a delicious and distracting theory. And while you are running through the flip book of your life story so far, trying to work out which best fits, we can get even more reductionist. A quote variously attributed to John Gardner and Leo Tolstoy says that in fact there are only two stories in the world – 'a man goes on a journey', and 'a stranger comes to town'. And even then, the difference between these two stories is only really a matter of perspective . . .

Chapter 2

The Angel in the Marble

Six Perspectives on Purpose

'Purpose' is without doubt a hot topic and a big buzzword for business. It is, however, poorly defined and often involves the conflation of at least six related but distinct ideas. So let's take a little wander through the territory of what we might mean when we talk about purpose, and why it matters.

First up is the **corporate responsibility school** of 'corporate purpose'. While corporate responsibility is sometimes regarded as a 'bolt on', 'purpose' is about an organisation's whole raison d'être – it gives responsibility a serious promotion, putting it at the heart of everything an organisation does and how it thinks. This is purpose entirely in terms of societal purpose, something normative and 'good'. It is about looking beyond short-term profit generation to positively impact individuals, society and the environment.[1]

This interpretation of purpose is a close cousin of a second, older concept which can be found in the Quaker-founded businesses of the past 200 years, and which is now enjoying a huge revival – namely, purpose as part and parcel of enlightened capitalism, incorporated within a broader and **longer-term definition of value generation**. This argument says that businesses with a clear sense of purpose generate more long-term value – i.e. are more sustainable – than those without.[2] Their positive societal impact strengthens the whole system upon which they depend, and their people are more engaged, innovative and productive.[3]

This perspective represents a correction and a moving on from the strain of free-market capitalism which actively opposed the role in purpose in business for most of the 20th century, embodied in Michael Friedman, who famously said:

There is one and only one social responsibility of business – to use its resources and engage in activities designed to increase its profits.

By way of just one example, Stuart Chambers, Chairman of Anglo American plc, said in his address at the company's AGM in May 2018:

> . . . So, when we set about delivering profitable growth in order to increase the value of your company through the cycle, it is not just what we do that matters or is good enough any more, if it ever was. . . . Many of you in this room care about – and have a right to expect – that we go about our business responsibly and sustainably – that we do the right things . . . If we don't behave in this way as an industry, we will lose our licence to operate and the business will become unsustainable . . . it is important that we think about what our purpose is as a business. Simply put, it is to re-imagine mining to improve people's lives.

This is a clear, powerful statement of the role of purpose in value creation from a notably returns-focused company, and a million miles away from Friedman, at least as he has come to be interpreted.

This longer-term perspective leads to a third **'start with why'** theory of purpose, as posited by Simon Sinek, via his 'Golden Circle',[4] which places 'why' an organisation exists at the centre, surrounded by the 'how' of how it conducts its business, and then places the actual 'what' of what the organisation does as the outermost of the three concentric circles.

Sinek's theory is that those organisations that follow this model – that is, those that are oriented in their thinking to start with *why* they exist, and then to think about *how* they do business, before they get to thinking about *what* they do – those are the organisations most likely to be resilient and successful.[5]

Fourthly, there is a perspective on purpose which sees the **evolutionary development** of organisational models in similar and parallel terms to the evolution of human consciousness, as explained through developmental psychology. This perspective says that as organisations have developed, over time, they have become less hierarchical and less concerned with power, more explicitly focused on their culture and more willing to consider the perspectives of multiple stakeholders. The next stage of organisational development, runs this perspective, is to turn attention to 'inner rightness' – *does this course of action seem right to me?* – and to strive for integrity and wholeness. This perspective – most persuasively presented by Frederic Laloux[6] – argues that a focus on purpose is both an inevitable result of an organisation's development and an active advantage, because the more complex and evolved our individual and corporate worldview and cognition, the more effectively we can problem-solve.

Still with me? Good, and hang on – we are turning inwards now, because the fifth perspective on purpose builds on this developmental perspective, but brings it back to the individual, and right into your brain. Neuroscience says that our default mode when our brains are at rest (i.e. not doing a task or processing sensory input) is that we are thinking about ourselves and our place in the world.[7] If this is a generally positive picture – if we are able to visualise a better future – then motivation and various associated tools kick in. Our 'big picture' processing networks are active and these trigger positive emotions. We can imagine the future because our visual processing networks and our capacity for perceptual imagery are triggered. In *Alive at Work*,[8] Daniel Cable further identifies **purpose as one of the triggers of engagement** for people at work, because it activates the brain's 'seeking system'.

In this thinking, purpose is individually conceived, rather than corporately, and is only partially related to the nature of the work in question, encompassing, too, the sense of purpose that people might feel by virtue of the impact they have, the difference they can make, and the relationships and interactions they participate in.

Finally there is a perspective on purpose which is primarily about individual **wellbeing**. This is related to the neuroscience perspective, but focuses more on outcomes. This perspective says that a sense of purpose brings a sense of coherence and significance and thus improves our health and increases our life expectancy.[9] In a work context, it also increases our intrinsic motivation, enthusiasm and resilience.[10] But this is only true – all these benefits only prevail – if it is your *own* purpose: a purpose that you have bought into and feel, not something that you have taken on board cognitively because somebody in your organisation told you that a particular cause was particularly worthy and wrote a glossy paper about it.[11]

Purpose Wash?

So far, so edifying, but where does all that that leave us? Are we all at sea in what Playfoot and Hall gloomily call 'purpose wash'[12]? In other words, is this all theoretical nonsense? I don't think so. I think there is a compelling unifying narrative which runs through all of these different perspectives, which is both philosophical and inspiring.

Backing up in reverse order, then: yes, an individual's purpose can theoretically be almost anything, but all the evidence suggests that it is commonly aspirational, and related to 'being part of something'.[13] The things that give us a sense of meaning and that activate the neurological conditions for engagement are not necessarily self-seeking, and do not necessarily even make us very happy. In fact, there is research to support the notion that people who self-report as having meaningful lives are often those who do things for others.[14] People like to galvanise around a common goal, or approach, or story that is bigger than themselves; this is an important aspect of wellbeing. Professor Julian Birkinshaw points out this in itself also directly creates value for firms because it inspires discretionary effort.

This in turn is consistent with the perspective that in terms of evolutionary development, it is both natural and advantageous for organisations to take a wider systemic view and to consider the 'inner rightness' of their activities within that context. And this chimes with Sinek's theory that companies that 'start with why' do better. All of these theories then support and drive the long-termist and wider societal perspectives favoured by those who approach purpose from a 'new capitalist' or 'corporate responsibility' perspective.

And ta-dah! Now we are edging towards a workable definition of corporate purpose as something that is commonly understood but individually owned, and which has some bearing on our impact and relationships, individually and corporately. As Quinn and Thakor say:[15]

> . . . when an authentic purpose permeates business strategy and decision making, the personal good and the collective good become one.

Harnessing the Power of Purpose

For myself, I am convinced by both the moral and commercial arguments which support the case for organisations discovering and committing to pursuing a purpose which is linked to an overtly good wider societal purpose. I believe this to be the best means by which to drive long-term value, provided that an authentic purpose, once found, needs to be consistently applied across every aspect of the business, and used to help organisations balance their short and long-term priorities. I like this quote from Punit Renjen, the Deloitte Global CEO:

> Purpose answers the critical questions of who a business is and why it exists beyond making a profit . . . But to be more than just words, purpose must guide behaviour, influence strategy, transcend leaders – and endure.

But in a sense, the case I am making in this book for the importance and relevance of purpose is a pragmatic one which says that whether or not you buy the notion that a purpose 'should' be any one thing or another, organisations that truly have an authentic purpose[16] have an advantage in a number of respects – connecting with customers, attracting talent and investment, building reputation, enhancing quality, fostering creativity, and improving relationships with stakeholders and regulators[17] . . . and in navigating and thriving throughout periods of change. For individuals, this book explores how purpose can engender a feeling of belonging, increase understanding and build energy and resilience. For organisations, it explores how a sense of purpose can help them to evolve, build confidence, remain agile, and simplify processes and bureaucracy.

In *The Master Strategist*[18] – a book which is less about strategy, and more about life, the universe and everything – Ketan Patel talks about how strategies for a higher purpose are transformational, capable of turning violence to peace, poverty to prosperity, enslavement to freedom. He explains that higher purpose can be built by creating a common aspiration amongst people: by rising above the ground of any conflict to find a higher, common position; by learning to 'take whole' – that is, to take the ground to be taken without being destructive; and by seeing any given event as belonging within a flow of events, and reacting accordingly.

These are lofty aims, but I have come to believe they are vital to the ongoing health and vibrancy of a changing organisation. What this all adds up to is a unified and compelling vision of purpose as both an end in itself, and as a tool – enabling businesses, their people and those around them to flourish.

Discovering Purpose

Clearly, though, it is futile for a leader to seek to simply decide upon and articulate a purpose at a corporate level and then instill it in others. Purpose is much better understood as something that is already there, inside your organisation and inside your people.

Sacha Romanovtich, the former CEO of Grant Thornton, talks about 'discovering' an organisation's purpose, and compares it to Michelangelo's famous (perhaps apocryphal) explanation when asked how he went about creating his statue of David:[19]

I saw the angel in the marble, and carved until I set him free.

We need to think in terms of discovering purpose, rather than in terms of imposing it.[20] This is a careful, rigorous and iterative process. To discover the angel, and to allow individuals to do so, Daniel Cable says you need to help your people to do two things – first, to witness their impact on others and, secondly, to develop their *own story* about the why of their work. We all can – and do – generate our own unique narrative based on a given set of common facts.[21]

The ultimate challenge for leaders and organisations at this point is to begin to curate, guide and weave these individual narratives together so that they form part of a wider, inclusive, corporate narrative that everyone can own and feel part of. This begins to explain the ineluctable link between purpose and story. Jason Barnwell, Assistant General Counsel for Legal Business, Operations and Strategy at Microsoft – who, as we shall see later, is a passionate advocate for the power of purpose in his teams – describes his own purpose in these terms:

My purpose at work is building upon serving the idea that great software can change the world for the better; great software is built and enabled by engaged, creative, empowered people; and I can build up those people and be one of them. I am far removed from the engineering discipline, but I believe my work services our larger mission. This helps me create meaning from my work, even the little things.

'Even the little things.' This is what purpose-work is about – not big, inspiring set-piece speeches once in a blue moon, but consistency of words and actions – the little things – day in, day out. The nature of a purpose-led approach is that the work is ongoing – you need to hold the purpose front and centre and use it to aid focus and guide decision making, even sometimes at short-term cost. Everyone in the organisation needs to tell stories about it over and over again so that it sinks into the collective consciousness of the organisation, gradually shifting the culture, changing thinking and performance, and allowing a new approach to processes and governance to evolve.

Notes

1. Playfoot, H., and Hall, R., *Purpose in Practice – Clarity, Authenticity and the Spectre of Purpose Wash*, Claremont Communications, 2015.
2. Porter, M., and Kramer, M., 'Creating Shred Value: Redefining Capitalism and the Role of the Corporation in Society', *Harvard Business Review*, 2011.
3. The Business Case for Purpose, Harvard Business Review Analytics Services, Harvard Business Review, 2015. Professor Julian Birkinshaw of London Business School, in the context of talking about why organisations continue to exist in a world where AI could theoretically navigate markets for us without the need for a company as 'middle man' argues that firms create value precisely by taking a long-term perspective, which machine struggle with, and by managing tensions between competing priorities – Professor Julian Birkinshaw, London Business School, in conversation at Temporall launch event, 6 June 2018, Institute of Directors, Pall Mall, London.
4. Sinek, Simon, *Start With Why*, Penguin, 2011. Sinek's model was originally conceived as a perspective on business development, but has much wider application.
5. Sinek's is ultimately a market-led perspective – an organisation 'needs' purpose and will derive value from it, if by having a purpose the organisation is better able to meet the expectations of its people, its customers and the markets in which it is doing business. This perspective basically takes the view that the system is shifting in favour of a more purpose-led perspective. An organisation at the sharp end of this movement can derive an advantage by meeting the expectations of the market better than its competitors can, but it cannot move unilaterally and independently of the system within which it is operating.
6. Laloux, Frederic, *Reinventing Organizations: A Guide to Creating Organizations Inspired by the Next Stage of Human Consciousness*, Nelson Parker, 2014.

7. For more on this, see the work of Ali Tisdall @ Mind3.

8. Cable, Daniel, M., *Alive at Work: The Neuroscience of Helping Your People Love What They Do*, Harvard Business Review Press, 2018.

9. Hill, P.L. and Turiano, N.A., 'Purpose in Life as a Predictor of Mortality Across Adulthood' *Psychological Science* 25(7): 1482–1486, 2014.

10. Cable, Daniel, M., *Alive at Work*, Harvard Business Review Press, 2018.

11. This is reminiscent of the Japanese concept of *ikigai* – the notion that a person's unique purpose is found deep within them, at the intersection of their profession, their vocation, their passion and their mission – Garcia, H., and Miralles, F., *Ikigai: The Japanese Secret to a Long and Happy Life*, Penguin, 2016.

12. Playfoot, H., and Hall, R., *Purpose in Practice – Clarity, Authenticity and the Spectre of Purpose Wash*, Claremont Communications, 2015.

13. Connected Commons, 'Network Investments that Create a Sense of Purpose in Your Work', *Connected Leadership*, www.connectedcommons.com.

14. Baumeister, Roy, Vohns, Kathleen, Aaker, Jennifer, Garbinsky, Emily, 'Some Key Differences Between A Happy Life And A Meaningful Life', *Journal of Positive Psychology*, 8(6): 505–516, 2013.

15. Quinn, Robert E., and Thakor, Anjan V., 'Creating a Purpose Driven Organisation: How To Get Employees To Bring Their Smarts And Energy To Work', *Harvard Business Review*, July/August, 78–85, 2018.

16. Purpose must be authentic in order to deliver the advantages noted, and only an authentic purpose is likely to be sustainable and viable within an organisation's wider societal and systemic context and to be 'useful' far beyond any particular project or target. That is to say, it will always be useful to have a purpose which inspires discretionary effort, encourages a longer-term view, helps with the balancing of competing priorities; it will always be advantageous – especially in the age of AI – to put human judgment and emotional conviction at the heart of your organisation.

17. For a further example, see the case study in relation to DTE Energy, as reported by Quinn and Thakor in the *Harvard Business Review* in July/August 2018: Quinn, Robert E., and Thakor, Anjan V., 'Creating a Purpose Driven Organisation: How To Get Employees To Bring Their Smarts And Energy To Work', *Harvard Business Review*, July/August, 78–85, 2018. And see also Daniel Coyne's observations in The Culture Code, in relation to the Meyer restaurant chain and in relation to Pixar. Coyne, Daniel, *The Culture Code: The Secrets of Highly Successful Groups*, Penguin, 2018.

18. Patel, Ketan, *The Master Strategist – Power, Purpose and Principle*, Arrow Books, Penguin Random House UK, 2005.

19. Sacha Romanovitch in conversation with Professor David Grayson, at Grant Thornton breakfast seminar, 27 March 2018.

20. John Coleman, in his HBR article 'You Don't Find Your Purpose – You Build It' takes issues with this idea of discovering purpose and argues that people need to create their purpose instead. To the extent that Coleman means that you should not sit around waiting for your purpose to descend suddenly upon you like a butterfly, I agree with him; I agree that you need to work to find your purpose, but I maintain that the work is more about unearthing than it is about creating (Coleman, John, 'You Don't Find Your Purpose – You Build It'6, *Harvard Business Review*, October 2017).

21. Fujita, K., et al, 'Construal Levels and Self Control', *Journal of Personality and Social Psychology*, 90 (2006), 351–367.

Chapter 3

Milk and Mushrooms

Around an organisation's purpose grows its culture. Culture hides in and permeates everything – history, artefacts, rituals, language and metaphors, behaviour, metrics, recognition – and for that reason is both hard to direct and incredibly potent.

Edgar Schein defines culture as:

A pattern of shared basic assumptions learned by a group as it solved its problems of external adaptation and internal integration . . . A product of joint learning.[1]

There's a reason, jokes Tristram Carfrae of Arup,[2] that culture is called culture:

like the stuff that grows on milk – it takes a while to grow, but then it stays for ever.

Others describe it as being like a thick, pervasive layer of mycelium, with the 'evidence' of the culture popping up like mushrooms all over the place.

Milk, mushrooms – part of the potency and value of an organisation's culture is that it is unique. It informs its brand and reputation, it cannot be stolen or replicated. A great culture increases productivity and engagement and so, ultimately, drives improved performance.[3]

And yet. While survey after survey shows that around 86% of CEOs consider culture to be 'important' or 'very important' in determining performance,[4] culture remains extraordinarily poorly defined or understood or managed. Part of the reason for this is that it is still a relatively new field of study (and of course, part of the reason for it being a relatively new field of study is that it is hard to define!).[5] The earliest and most definitive academic study of organisational culture was published by Edgar Schein in the 1980s. In his book, now in its fifth edition,[6] Schein does a compelling and deep dive into what culture is, how it affects the organisation, how to understand and decipher it and how to act on it or change it.

Enlightenment is out there, then, but we do not yet generally see business leaders equipped with the depth of understanding and mastery in this field necessary to enable them to take hold of Schein's (or anyone else's) insights and use them to actively shape, manage and embed a culture in their organisation that drives value and is sustainable. Part of the difficulty is due to the relatively long-term perspective required, and part is due to the challenge of correlating inputs and outcomes and, ultimately, impact. Nathan Adams, HR Director at Aviva, the UK's biggest insurance company, talks about the need to '*hack*' the culture; to get deeply inside it and influence it from the inside if it is to be harnessed to drive accelerated performance. For this to happen, culture needs to be at the heart of strategy, not an adjunct to it, and it needs to flow through into behaviour, processes and systems. And if it is ever to truly become a board imperative, its impact also needs to be measured. All of this is only possible if purpose is understood, and grappled with, by the senior management team.

Thomas Davies of Temporall[7] reckons that he and his team have figured how to measure the ROI of culture; the correlation between culture and individual and organisational performance. The team has benchmarked outstanding company performance and correlated it with certain key aspects of organisational culture. This enables them to give organisations a Cultural Performance Score (i.e. a CPS, to sit alongside NPS), to identify gaps and suggest concrete actions to address them, and then to measure progress and outputs based on a number of KPIs. This will certainly get the C-suite's attention.

While an organisation's particular purpose and market context will determine the culture that is exactly 'right' for it, or for a particular part of it, there do also appear to be strands of culture which are common to many purpose-led organisations and which also correlate to high performance: high levels of trust, transparency and accountability, for example. Wholeness. A growth mindset.

Know Thyself

But while there are common threads, there are also differences, and with culture you never get to start with a blank piece of paper – there is always a culture there already, and the trick is to understand it and work within it – both to leverage its strengths and to influence it to shift.

There are diagnostic tools, including Temporall's, which can help with this, and there are also a number of different perspectives one can adopt in seeking to understand culture. On the one hand, the culture in an organisation is simply the aggregate impact of every individual in the organisation, and so seeking to understand individuals and what motivates them is one route in. At the other end of the spectrum, culture is influenced heavily by the national culture within which an organisation is operating. There may also be particular traits which are common to the particular profession or sector that your organisation operates within which will need to be considered. This is certainly true of lawyers, and as you will see in Part Two, while it is important to be wary of over-generalisations, understanding how lawyers and those who work alongside them think and feel and behave was important in enabling us to successfully navigate the challenges of the CMS merger.

Say What You See

Organisations, then, need to be culture-literate if they are to understand how their organisation really works and how to influence for change. They also need to be culture-articulate if they are to stand a chance of explaining an organisation to itself and of helping it to navigate periods of change.

Often one of the most overt and visible manifestations of an organisa-tion's culture is a statement of their shared values – essentially an articula-tion of the 'how' layer in Sinek's Golden Circle – *this is how we do what we do*. At their worst, these statements, however carefully mined from across the various strata of the organisation, suffer from the strange deadening hand of corporate jargon and make absolutely no difference at all to how anyone behaves, how any process operates, or how any decision is made. Abstract nouns, denuded of meaning and context, languishing on a lami-nated poster next to a picture of a mountaineer . . .

Teamwork. Commitment. Excellence.

Argh. We have values in our family, and you do too, and yet no one feels the need to write them down. They are made manifest in how we show up every day, the way we speak to one another, the things we do. Perhaps it is hard to imagine how to get to this level of shared and deeply embed-ded assumption in an organisation, and perhaps there should always be a more overt statement of values at hand. But as a starting point, at the very least, it is in my experience better by far to express those same con-cepts in terms of active behaviour – *'we include each other'*, *'we collaborate'*, *'we do excellent work'*, *'we respect each other'* – and better still to stitch them through everything the organisation does, and to create and tell stories which bring these behaviours to life, to instruct and inspire.

Perhaps best of all is to 'act as if'. The clarity that an agreed set of behaviours brings provides the best possible platform for this. In his book *Turn The Ship Around!*, David Marquet, former captain of the USS *Santa Fe*, talks about having identified some of the key behaviours that he would expect to see if the culture he and others were working towards were made manifest on the *Santa Fe* under his command. In relation to some of these, he then simply insisted on the behaviours, trusting that in this way the crew would act their way to new thinking.[8]

In the notably less-hierarchical culture of professional services firms, I have found that a productive equivalent approach is to explicitly reference the behaviours that a team committed to encouraging alongside the many actions that are likely already taking place across the business that makes these behaviours manifest. This raises awareness and understanding and creates a virtuous circle: an organisation which is rooted in its purpose and has compelling stories which demonstrate that purpose can increasingly act in line with that purpose and create and tell more stories as a result. These then further strengthen the narrative and enable yet more change to happen. This approach will also help to ensure integrity as between the values on the wall and the behaviour on the ground, which is critically important.[9]

Notes

1. Schein, Edgar H., *Organizational Culture and Leadership*, 5th edn (The Jossey-Bass Business & Management Series), Wiley, 2016.
2. In conversation between Jenni Emery and Tristram Carfrae, Deputy Chairman, Arup, 18 May 2018.
3. Heskett, James, 'Putting the Service-Profit Chain to Work', *Harvard Business Review*, 2008.
4. See, for example, Deloitte's Global Human Capital Trends Survey, 2016.
5. The first forays into studying organizational culture came in response to the entry into US markets of Japanese car manufacturers. The US incumbents observed the dramatic impact that the culture of these new entrants appeared to have on their performance and wanted to understand more. Thomas Davis of Temporall, in conversation at Temporall launch event, 6 June 2018, Institute of Directors, Pall Mall, London.
6. Schein, Edgar H., *Organizational Culture and Leadership*, 5th edn (The Jossey-Bass Business & Management Series), Wiley, 2016.
7. www.temporall.com.
8. Marquet, L. David, *Turn the Ship Around! A True Story Of Turning Followers Into Leaders*, Penguin, 2015.
9. MacLeod, D., and Clarke, N., *Engaging for Success: Enhancing Performance Through Employee Engagement*, Department for Business, Innovation and Skills, UK, 2009.

Chapter 4

Telling Stories

Enough of these phrases, conceits and metaphors.
I want burning, burning, burning.

Rumi[1]

In his brilliant book, *Be More Pirate*, Sam Coniff Allende sets out a model for change which has storytelling at its heart. Drawing on lessons from the 'Golden Age of Piracy' – in the late seventeenth and early eighteenth century – Coniff Allende explains how pirates found a cause – a purpose, a reason to rebel – and then created something new instead; they rewrote the rules. Pirates then reorganised themselves so that they were faster and stronger than their opponents and then deliberately redistributed power and decision making so that their purpose could be pursued without the risk of bureaucracy or corruption. Then, last but by no means least, pirates told stories about themselves in order to get attention and further pursue their purpose. Coniff Allende says that pirates 'creatively weaponised' the art of storytelling – skulls and crossbones and bleeding hearts on black flags, weaving fierce myths right into the heart of the establishment. Coniff Allende quotes 'Calico' Jack Rackham – one of the Golden Age pirates:

A story is true. A story is untrue. As time extends, it matters less and less. The stories we want to believe . . . those are the ones that survive.

Now, lawyers are not pirates. But they are, like many of us, autonomous, independent thinkers. They like hard facts and have a profound integrity. They are cynical about obvious attempts to influence or coerce them to think x or to do y . . . fine rhetoric gets marked up in red pen, spin doctoring and flagrant attempts at bribery are just ignored. Stories, however – good stories – work on us in a different way and, as we shall see, I found their impact to be transformational when they were used as part of the CMS merger process.

So, what is it about stories? What gives them their power? Three hundred years after the pirates, Daniel Coyle gives us an answer from neuroscience:[2]

We tend to use the word *story* casually, as if stories and narratives were ephemeral decorations for some unchanging underlying reality.

The deeper neurological truth is that stories do not cloak reality but create it, triggering cascades of perception and motivation.

So, the pirates' stories did more than just communicate a message – they actively helped the pirates to achieve their purposes by scaring people silly and so minimising the likelihood of actual conflict. The stories themselves become part of the action. Stories are not neutral then, and nor are they innocent. And yet stories are also subtle. They are dynamic and spacious and people – pirates and lawyers alike – can own them, develop them and make them their own. This makes them a respectful means of communication. They allow for the listener's intelligence and perspective, and leave space for interpretation and nuance.

One of the key differences between a series of facts of the sort that lawyers so love and a story is the addition of context, emotional content and sensory detail – what Aristotle called *pathos*.[3] These can change everything. In the introduction to her wonderful book *The Story Factor*,[4] Annette Simmons says that trying to explain storytelling is like trying to explain a kitten. You do not – cannot – take a kitten apart to in order rationalise and understand the essence of its kittenyness . . .

Instead, we need to make a little room for both of the sacred and the animal here – something both before and beyond the rational. In their capacity to influence, challenge and engage us, stories steer even the most logic-loving of us beyond the empirical and around our tendency towards confirmation bias[5] and engage with a different part of us. Jerome Bruner says:[6]

. . . for all that narrative is one of our evident delights, it is serious business. For better or worse, it is our preferred, perhaps even our obligatory medium for expressing human aspirations and their vicissitudes, our own and those of others. Our stories also impose a structure, a compelling reality on what we experience, even a philosophical stance.

There's an apocryphal little story – it's hard to identify its origin – about three men, working away hard on a hot day, surrounded by bricks, sand, cement.

The narrator-storyteller asks the first man, 'What are you doing?'

'I am laying bricks,' he says.

The narrator asks the second man, 'What are you doing?'

'I am building a wall,' he replies.

The narrator asks the third man, 'What are you doing?'

'I am building a cathedral,' he replies.

This little anecdote illustrates the power of a story to convey vision and instill ambition and engagement. Stories can paint a vision of the future. They can be about 'who we are' – individually, and as an organisation– and can show us what we value much more effectively than a laminated poster.

Stories also give context. They can help to explain strategies, and relative priorities. They can show us what matters, who matters and what success looks and feels like.

But stories, well chosen and well told, can also have a kick in them. They can call people out on their lack of alignment or commitment, or name an unresolved issue. Provided stories are authentic and honest, they can disarm listeners and help to reframe challenges.

There is a teaching story in the Talmud about Truth wandering around a village late at night, naked and cold. Nobody will take her in. Eventually,

Parable comes along and clothes Truth in story, at which point Truth is no longer turned away, but is welcomed in wherever she goes and given food and shelter.

All of this means that stories have a degree of power. Annette Simmons conjures Arthur holding Excalibur: at times of change, if you have a good story and hold it high, and tell it well, you hold in your hands the power to pull people together for a common cause, and to help them navigate change.

What Makes a Good Story?

I have hesitated about including this section because it risks being limiting and prescriptive. At the end of the day a 'good story' is one that works – that engages you, or moves you, teaches you something or makes you think differently. The vision which runs through this book is of hundreds of thousands of stories – some short, some long, some funny, some sad – all bundled together in no particular order in a crazy, loose-leaf compendium to bear witness to, and yet also hold together and drive forward, an organisation and the people in it during a period of change.

All that said, there are stories and there are stories, and so let's touch on a little bit of what we know about what makes a good story – if your intent in the telling the story is to build change-readiness, and thinking here about stories at the overall organisational level, rather than the local or individual level. Good stories for these purposes address five areas:[7]

- They start by talking about how the world is changing – maybe there's a burning platform, or maybe there's a golden egg, or maybe both. At CMS, for example, the leadership team talked about changes in clients' needs and expectations, about the opportunities presented by technology and about shifts in the legal market.

- Then, good stories explain 'what we're going to do about it' – we're going to create a distinctively different future-facing of law firm . . . –

- . . . and why . . . to better serve our clients . . .

- . . . and how . . . we're going to merge three great firms.

- Next, good stories emphasise commitment – in terms of allocation of time, resources, expertise and money, but also in terms of authentic and dedicated sleeves-up leadership from the top.

- Finally, good stories talk about why it is worth all the effort – not only for the organisation, but also for its customers, individuals in the business, teams and business units, and for society as a whole. It is a good idea to touch on all of these stakeholders at this point, because different people are engaged and motivated by different aspects of this.

Notes

1. attr. Rumi, Mathrani, *Moses and the Shepherd, Teachings of Rumi*, trans. Whinfield, The Last Library, 2013.
2. Coyle, Daniel, *The Culture Code: The Secrets Of Highly Successful Groups*, Penguin, 2018.
3. The very etymology of the word 'narrate' has tangled up in it two separate concepts – *'telling'*, and *'to know in a particular way'*.
4. Simmons, Annette. *The Story Factor: Inspiration, Influence and Persuasion Through The Art Of Storytelling*, Perseus Publishing, 2001.
5. Our tendency to select only those facts that help us to confirm what we already think we know.
6. Bruner, Jerome, *Making Stories: Law, Literature, Life*, Harvard University Press, 2003.
7. Armenakis, A., Harris, S., Mossholder, K., 'Creating Readiness for Organisational Change', *Human Relations*, Vol. 66, Issue 6, pp681–703, 1993.

Chapter 5

Everything Must Change

It is virtually impossible to talk about change without talking about John Kotter, and his iconic book, *Leading Change*.[1] It would be a brave or foolish person who would challenge what has become, in the twenty years since its publication, pretty much the universally accepted and definitive model for managing change. Happily, I am not that person. My response to Kotter is simply, 'Yes. And . . .'

Kotter outlines an eight-stage process for change:

- establishing a sense of urgency
- creating a guiding coalition
- developing a vision and strategy
- communicating the change vision
- empowering employees for broad-based action
- generating short-term wins
- consolidating gains and producing more change
- anchoring new approaches in the culture.

In this list, Kotter has identified and elegantly defined eight important aspects of any change programme, and it can be helpful to think of them, at least initially, as a linear process (Fig. 5.1). However, the process as outlined assumes a degree of hierarchy and a degree of command and control, which is increasingly rare in organisations today, as the influence of millennials in the workforce, the increased reliance on technology and the sharing economy all drive us towards a more level, inclusive and collaborative field of leadership.

In organisations with a flatter hierarchy and a more consensual approach to decision making – and particularly when the change in question is wide-ranging and affects a large number of people in significant and disparate ways – I believe that the process is more iterative than Kotter suggests (Fig. 5.2). It is more like a spiral, or a helix of interwoven threads. And

Figure 5.1 Kotter's original linear approach.

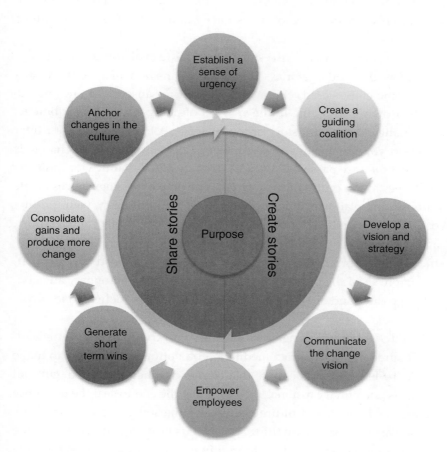

Figure 5.2 A more iterative take on Kotter's model, with storytelling as the key driver, and purpose at the centre.

in every segment or twist, being able to tell a story about your purpose and values will help to catalyse, lubricate and generally ease the turning.

So, from the top, where Kotter starts with creating a sense of urgency, I would start with a story. Your guiding coalition – your leaders – have a role to play in both developing and telling that story, and as it puts on weight, it becomes your vision and strategy, and all the riper for the telling.

As you communicate it, the story continues to evolve and develop, and to plant seeds of other stories. This is the means by which your employees become empowered – because they start to sense-make and then to tell their own stories and, in the very act of doing so, they contribute to, even change, the overall story of the business. A linear reading of Kotter may lead you to conclude that employees are those to whom change is *done*; that they are passive. This more iterative and dynamic view puts the employees at the centre of the process. The employees themselves create your burning platform, they become your sense of urgency, your guiding coalition, your agents of change. In a 2012 critique of Kotter's model drawing on more recent change research, Applebaum et al.[2] clearly demonstrated the importance of this stage five – 'empower broad-based action', and emphasised the importance of granting as much autonomy and control as possible and allowing change to be owned and delivered locally.

Finally, Kotter warns against 'starting' with trying to change the culture. I agree that an abstract meditation on culture which aims to make people 'be less this' or 'be more that' seems destined to fail. But cultural change is not a final capstone, to be added at the end to seal the deal once change has happened. Culture is alive. It's the soil you are standing in. As soon as you change anything in your business, you are changing the culture. As soon as you change the culture, you are impacting everything that is rooted in it – that is, your whole business. Everything is connected.

Find the stories that already exist in your culture upon which you can build a case for change. Better, find the stories that already exist in your culture which can help you discern what change to make in the first place. Then, plant the very story of your changing into your culture. And as the change takes root and new things grow, tell stories about those too. These things are part of your purpose and values, and in the telling become part of your culture, and so it goes, and so it grows.

Having contended with Kotter, let's touch briefly on mergers and acquisitions, the T-Rex of the change world, and on the particular challenges and wrong thinking that seem to beset this area, before going on to consider some of the behavioural science and neuroscience which underpins what happens during periods of change.

Mergers and Acquisitions

As discussed in the introduction, the staggering failure rate of mergers should give us particular pause. Time and again, post-mortems reveal a chronic lack of attention, focus and investment in the people and cultural aspects. This neglect is also evident in – and perpetuated by – much of the 'how to' literature that is out there. For example, in *M&A Integration: How to Do It – Planning and Delivering M&A Integration for Business Success*,[3] Danny A. Davis delivers a blow-by-blow, step-by-step guide to integrating businesses following merger.[4] Davis acknowledges the human angle, warning that during an integration period we should actively expect chaos, political manoeuvring and no thanks:

> 'Everyone', he says, 'who has been through a merger (not run it) will always say that it was done poorly and could have been done better.'

While his calling-out of a familiar sentiment may be a source of scant comfort in the darker hours of an integration project, it casts people in the role of victim – almost collateral damage – people to whom a merger is 'done', rather than people who are part of a team, working together to make change happen.

What's more, Davis's only advice in the face of this criticism is to be head-down, well-organised, and to crack on. Again, this casts the whole process of change as something to be tolerated and lived through, rather than as a process which is valuable in itself – the coming together of values, the exploration of common purpose, the discovery and creation of stories.

Davis ends his book with a list of his 'Killer Insights for Integration'. Plan, plan, plan and communicate, communicate, communicate. Teach, define success, show your working. You need, he says, continuity of strategic thinking. And you need stories. So far, so good. But then:

'Culture', says Davis, 'is always used as an excuse, but this should not put us off a good financial deal. Yes, culture is important, but it is not a deal breaker and should not be used as an excuse for failure'.

It perhaps already goes without saying that I could not disagree more with Davis on this fundamental point. Culture can be THE difference between success and failure. Culture is everywhere, and it influences everything. It is the bedrock of your strategy, employee engagement, client satisfaction and so, ultimately, value in the hands of your partners or shareholders. It is short-termist at best to believe that there is a good financial deal to be made when cultural fit is not there, particularly if your objective is to build a sustainable, people-centred business.[5]

In a 2011 study[6] which surveyed more than 200 senior executives worldwide who had experienced M&A deals over the previous three years, more than two-thirds of the business leaders surveyed said they had over-emphasised traditional due diligence (on tangible assets such as finance and IT) at the expense of intangible assets (such as human capital, leadership compatibility and cultural integration), and only 13% said that they had prioritised engaging and integrating senior management and the workforce once the M&A had been completed.

Deborah Allday, an independent M&A consultant who led that 2011 study, spoke to me about the 'building blocks for failure' that she continues to see in her M&A consulting practice. Over the past 20 years, Allday has participated in dozens of deals worldwide across the energy, mining, manufacturing, finance and professional services sectors, and has conducted and published leading-edge research into the key factors that create or destroy transaction value. She identifies three big building blocks for failure that arise time after time – first, a lack of internal alignment amongst senior executives on both sides of the deal, sometimes even as to the whole rationale for the transaction, and then as to the priorities and areas for focus; secondly, a failure to properly identify and address potential incompatibilities of 'intangible capital', such as governance, brand values, client management, and organisation culture; and thirdly, disaffected, unsupportive and unsupported managers, as part of the wider leadership team. We talked about this first area in particular. Allday says:

Both parties to any deal absolutely need to ensure that their respective executive leadership teams are completely aligned around deal rationale, deal benefits and dangers, and how the key make-or-break synergies will be delivered – talking and walking the same talk and walk. Both sides should be using due diligence downtime to align their own executives and engage their senior management teams as widely as possible, given regulatory constraints. You need visible leadership from the front, all the senior people really engaged and focused. As early as possible in the process, you need to get people working together across the deal, too. Not only to hammer out the detail of integration, but also to build the foundation for relationships that will carry the integration forward. Otherwise, you risk sabotage.

Throughout the whole of our wide-ranging conversation, it is this point above all that Allday emphasises repeatedly as being critical for success, drawing on all of her years of wide experience – this idea of bringing the influential people across the combined business together, and enabling them to bond, building a sense of identify and common purpose. She talks about successful merger integration processes led by leadership teams who have bonded strongly because they spent time early on getting to know each other as individuals, sharing perspectives and building trust.

'Do whatever it takes', says Allday. 'Everything else tends to sort itself out if you truly align the executive team and support them in achieving a single, unified culture as quickly as possible. This is how you deliver profitable M&A deals.'

Change – This Time It's Personal

So, what might it look like to bring a truly people-centric approach into the heart of change theory for businesses? In bringing a purpose-centred, story-rich perspective to bear, what we are really doing is quite explicitly bringing some of the latest thinking in psychology and neurology into a business context. Curiosity, vulnerability, pleasure, pain . . . it pays to remember that we are all human, and will continue to behave accordingly, even when we are suited and booted, and sitting inside glass offices and multi-million turnover businesses. If we are to build a new theoretical framework for change then it must, as far as possible, be consistent with, and draw on, the best of what we know about how people behave and how the brain works.

Perhaps the most widely known model for change on a personal level is the Kübler-Ross model, first produced in the 1960s as a model for how people process and move through grief, and later widely adopted as a model to map the stages of periods of change and transition more generally. This model is commonly called the 'Change Curve', and while it can be useful in helping individuals and leaders to frame and articulate their varying reactions across a period of change, apparently even Kübler-Ross herself did not in the end believe the phases to be sequential in the neat way they have come to be understood, and nor did she intend us to – quite literally – pathologise a normal, healthy, change process by equating it with grief.

Traditional ways of thinking about organisational change see it as a process or a programme, with a beginning, a middle and an end. They promise jam in the steady-set nirvana that is tomorrow, in exchange for a little pain now. But this is not how our volatile, uncertain, complex and ambiguous world works, and if we carry on trying to motivate people on the basis of false promises and false limitations, we are destined to fail. Change is our new steady state and we should be changing the whole paradigm such that we learn to accept, embrace, thrive on and even to drive change as something we are wired for, and something we actively seek.

More recent developments in our understanding of how the brain works can help us much more here. As with all things neurological, we are still only in the foothills of unmapped territory. One thing that we are increasingly understanding, though, is that our brains are malleable. New connections can be made, new neural pathways laid down – this is called neuroplasticity. Our brains are built to do this – it is natural and healthy and the brain can do it relatively easily and over a short timeframe, *provided* – and here's the rub – the brain's owner makes sufficient effort to focus attention on the change it wants to make. Attention is paramount. It's that simple – and that inordinately difficult.

The neuroscientist and author, David Rock, in *Your Brain at Work* says that three conditions are necessary to persuade someone – and, by extension, whole organisations of someones – to shift their attention and thus change their thinking and behaviour:

- First, it needs to be safe for them to do so. You, yes you, swanning in here with your big new idea, are a threat. We need to minimise that, and enable people to feel safe to make change happen themselves.

- Secondly, we need to help people to focus their attention in just the right way to create the right new connections.

- Thirdly, people need a reason to keep coming back and walking that new neural pathway again and again so that it becomes the thoroughfare of choice, not an overgrown and disused path.

Rock uses what he calls the SCARF model to summarise five social domains that drive human behaviour and are instrumental in determining how threatened or safe an individual feels. The five domains are Status, Certainty, Autonomy, Relatedness and Fairness. If, in order to make change happen, our first aim is to increase safety and minimise threat, then we need to pay close attention to all five of these domains. Once a threat response – commonly known as the 'fight or flight' response – is triggered in the brain, a number of things happen which make any change process even more challenging. Blood flows away from our prefrontal cortex, which is where we do our rational thinking, and manage our emotions, and make plans, and flows to our amygdala to help us get ready to run or smack someone. Our anxiety increases and we start to box at shadows and see threats that are not there. All of this makes us struggle with decision making, creativity and problem solving. We also collaborate less.

It is a commercial imperative to do all we can to keep people performing at their best during times of uncertainty. Paying attention, then, to Rock's five domains is important, and there are many specific examples throughout this book of how preserving a person's status or increasing their certainty, for example, can make a profound difference to their capacity to perform during periods of change. For now, it is worth reflecting on how an approach which dispenses with hierarchy, puts purpose at the centre, and encourages individuals to create their own stories and purpose within the wider story and purpose, will favourably impact at least the first four of these domains.

Once you have reduced the threat sufficiently that people have offered their attention, the next challenge is to help them focus that attention in the right way. Rock says that one of the more effective strategies for doing this is to tell a story. He calls a good story an 'insight delivery device' – it creates complex maps in the brain because it requires people to hold different characters, events and relationships on the stage, and in so doing delivers the story's 'point' effectively. However, the risk with simply 'telling a story' is that people are not daft – they know when they are being played and may end up feeling threatened and unsafe all over again. An alternative, or complementary, strategy, says Rock, is to ask people a question and so draw them in to the narrative themselves – exactly what is advocated in this more dynamic model for change – tell a story, and let people add to it and refine it as it is told.

The final step in Rock's approach is to ensure that people keep their new circuitry alive. This means finding a multitude of ways to increase the 'attention density' around that new idea. This is particularly important in a context which demands adaptive change (i.e. a change of mindset and outlook) as well as technical change (i.e. development of new skills).[7] Talking about the change, telling more stories, asking people to share ideas and collaborate – all of these things will help, and all are implied in the collective and iterative approach to change which is advocated for here.

The case for this approach to change is further supported by the latest thinking in neuroscience about what it takes to keep us engaged and motivated at work more generally. Daniel Cable's central thesis[8] is that organisations can, and should, keep their employees engaged – *alive* – at work by activating their 'seeking systems'; the parts of our brain that encourage us to explore and learn. When we activate this system, it releases dopamine – a neurotransmitter linked to motivation and pleasure – and this release in turn makes us want to explore and learn more.

There are three triggers that activate the seeking system – self-expression, experimentation and purpose. It gets even better than this, because we know that the very process of engaging with others to understand and tackle a problem is a significant source of purpose for many people.[9] It is clear, then, that an approach to change which includes people in the crafting of the journey itself, and allows them to create their own stories and discover their own purpose, will keep people much more engaged and motivated than a process which sees them as passive recipients – or victims – of change.[10]

This chimes with the work of Dan Pink, who identifies autonomy, mastery and purpose as the three key intrinsic motivators of our behaviour,[11] and with the work of Amabile and Kramer,[12] which concludes that the single biggest motivator is a sense of progress – the capacity to make small wins and make sense of those in terms of progress towards an ultimate end or a bigger purpose will impact on both productivity and creativity. Actively engaging people in co-creating and directing their change journey, rather just having them sit and wait for it to happen to them, not only mitigates their threat response but actively engages and motivates them.[13]

Turning Theory into Practice

So, there's the theory. This Part One of the book has set out a little of the history and importance of storytelling. It has strived towards a unifying definition of purpose, and then explained the important roles of purpose and storytelling in the context of change. It has proposed a new model for change management that incorporates these two key components, and explained a little about how our brains work during periods of change. Before we get stuck into considering the practical implications of all this theory in Part Three, Part Two illustrates how those theories can play out in practice, by sharing some of the stories, observations and insights that I gleaned by virtue of my involvement in the remarkable and inspiring story of the merger in May 2017 of CMS, Nabarro and Olswang.

Notes

1. Kotter, John. *Leading Change*, HBS Press, 1996.
2. Applebaum, S., Malo, J., Habashy, S., Shafiq, H., 'Back To The Future: Revisiting Kotter's 1996 Change Model', *Journal of Management Development* 31(8): 764–782, August 2012.
3. Davis, Danny A. M&S Integration – How to Do it. Planning and Delivering M&A Integration for Business Success, Wiley, 2012.
4. If you want that sort of practical help, stop reading this book now and read his instead – it proved invaluable to the team at CMS in the weeks before, during and after our own merger.
5. Plucking just one example from the air – the initial difficulties in the Safeway – Morrisons merger have, with hindsight, been attributed to issues around cultural fit. See https://laffinsdotnet.wordpress.com/2017/02/06/sir-ken-morrison-a-retail-giant-but-still-human/.
6. Hay Group, Touching the Intangible, 2011.
7. Kegan, Robert, Laskow Lahey, Lisa. *Immunity to Change: How to Overcome it and Unlock the Potential in Yourself and Your Organisation.* Harvard Business Review Press, 2009.
8. Cable, Daniel, M., *Alive at Work: The Neuroscience of Helping Your People Love What They Do*, Harvard Business Review Press, 2018.
9. Connected Commons, 'Network Investments that Create a Sense of Purpose in Your Work', *Connected Leadership*, www.connectedcommons.com.
10. This chimes with Deci and Ryan's work around the three key components of motivation being autonomy, competence and connectedness. (Deci, E.L. and Ryan, R.M., 'The Importance Of Universal Psychological Needs For Understanding Motivation In The Workplace', in Gagne, M. (ed.), *Oxford Handbook Of Work Engagement, Motivation and Self-Determination Theory*, Oxford University Press, 2014).
11. Pink, Daniel, *Drive: The Surprising Truth About What Motivates Us*, Canongate, 2009.
12. Amabile, Teresa and Kramer, Steven, *The Progress Principle: Using Small Wins To Ignite Joy, Engagement And Creativity At Work*, Harvard Business Review Press, 2011.

13. This chimes with the findings of the MacLeod Report, which, though not specifically focused on change, explores the link between employee engagement and organisational performance. The report identifies four critical factors for high employee engagement – visible, empowering leadership that provides a strong strategic narrative; engaging managers who give people scope to develop and treat them as individuals; employee voice throughout the organisation and integrity between espoused values and actual behaviour. MacLeod, D., and Clarke, N., *Engaging for Success: Enhancing Performance Through Employee Engagement*, Department for Business, Innovation and Skills, UK, 2009.

Part Two

Story

I saw that my life was a vast glowing empty page and I could do anything I wanted.

Jack Kerouac[1]

Note

1. Kerouac, Jack, *The Dharma Bums*, 1958.

Chapter 6

Beginnings

Once Upon a Time

This is a story about a merger. This means that it is a story about business, perhaps a story that might be told, by a different storyteller, in numbers on a spreadsheet – a barrage of forecasts and graphs trending up and right, or in a flowchart as a series of logical steps.

Different books could be written about the detailed strategic rationale for the merger, the benefits it afforded clients, the careful and empowering terms upon which it was negotiated, and the market backdrop against which it took place, but those stories are for others to tell.

The story I want to tell is about human beings – about the great things they can do together, how they think, how they work, what happens under pressure, and what happens when everything changes. I want to talk about the underlying purpose of the firm that was being created, the stories that were being told, and how they all came together: the human aspects, if you will, of merger and integration.

This is a story about endings and beginnings, hard work, sacrifice and growth. It is 'a man goes on a journey' and 'a stranger comes to town'.

It is always compelling to begin a story well. One of my favourite beginnings is Gabriel Garcia Marquez's opening sentence to his novel *One Hundred Years of Solitude*:

> Many years later, as he faced the firing squad, Colonel Aureliano Buendia was to remember that distant afternoon when his father took him to discover ice.

Wow. The way it brings the ending right to the beginning, the way it looks backwards and forwards at the same time, the way it conjures one dramatic moment in time, while also evoking a gentler moment long ago, the brilliant name of a key character, the presence of the father and family and history, the innocent yet sensual detail of ice, first discovered.

The beginning of the CMS merger story, although less exquisitely lyrical, is equally multi-faceted, and it is important to begin here in order to fully understand everything that follows. Each of the three players in the story that is about to unfold had a proud history, great market standing, and a clear sense of its strategic priorities, its position in the market, and its ambitions for the future. Each had months of rigorous research and debate under its belt. Now, of course, all the best stories have a little sprinkling of fate – sliding doors, a plane boarded or missed, a chance encounter in the street – and this one does too, but it is ultimately rooted in clear thinking, hard work and the glimpse of a shared vision for the future.

From here, the story begins to unfold. As it does, there will inevitably be myriad details and nuances to consider, but right from the outset, the three firms agree that the paramount consideration is whether the proposed deal will deliver a better result for clients, and that fit – both strategic and cultural – is therefore absolutely critical.

First Meeting

And so, after weeks of research and analysis and due diligence to prepare the ground, the first official meetings of the firms' leadership teams took place socially, over dinner, in a discreet hotel in the City of London. Dinner went well. If cultural fit can be measured by the volume of laughter in a room, and how hard it is to interrupt discussion with a fork chiming on a glass, then it was a roaring success. People swapped places between courses, and alongside in-depth conversations about business areas and potential synergies, everyone also tried hard to remember the names of others' spouses, children, dogs. Because I already knew my brief was to focus on culture and people, I watched constantly, closely, to make sure all the important pairings worked. I made mental note after mental note of things to check, follow up on or worry about. When I fell off the end of my memory, I scribbled biro notes up my arm. Overall, it felt good. If it had been a date, you would want the phone to ring.

Consider, already, how many stories are in the room. Past, present and future. At least as many stories as there are people. Lots of ideas, lots of questions, lots of pride, hopes, alliances, histories . . . How to even begin to figure out whether this is a deal that can, and should, be done? How to work out what clients will make of it all? How to understand each firm's perspective? Each leader's perspective? How to paint a picture of the future together? The histories of each firm alone could fill a library.

Lawyers, Generally

While we are meeting the players in this story, it is worth adding some brief comment on the nature of lawyers generally. This is important because it speaks to culture and brings some context to much of what follows.

Lawyers are amongst my favourite people in the world. I was one, I love one, I have worked amongst them for the best part of twenty years. Their focus on their clients, their commitment to quality, the strength of their moral compass, their intellectual firepower, their skepticism, rigour and love of a good debate all help to make lawyers terrific people to share one's life with. These very same characteristics can also contribute to their sometimes being amongst the least happy of people,[1] in part because of precisely the traits that we select for and then further train in – spotting the catch, being analytical, managing risks, focusing on details, and working with high urgency to tight deadlines.[2]

It is fair to say that this book paints a positive and optimistic picture. It says that with clarity of purpose and a compelling narrative, organisations can successfully navigate change, and indeed flourish through it – building a strong sense of belonging among energised, confident people who understand one another, can evolve and be agile as the circumstances require and can tackle complexity with simplicity. But if this all serves to conjure up in your mind a sea of smiling, biddable faces, you would be mistaken – the rigour and challenge that people across all three firms brought to bear throughout the whole change process was remarkable, and the CMS merger was the better for it, as we shall see.

Digging Deeper

So. After those initial dinner dates, meetings were arranged between leaders in complementary industry sector or practice groups to enable the counterparts to dig deeper with one another and explore what a merger might look like in their area. In marked contrast to the informal dinners, the team approached these meetings in a careful and considered way – there was a written agenda and a request to prepare some information in advance, and a member of the core merger team sat in on all meetings to take notes, listen out for any key themes emerging in across different meetings, ask any clarificatory questions, and keep the discussions broadly on track. This was important at this point for a number of reasons.

First, of course, was the imperative to ensure that client confidentiality was maintained at all times and that everyone complied with their regulatory obligations, and so the team was crystal clear at the outset about what information could and could not be shared. While everyone understood this and complied, it was nevertheless frustrating. What lawyers love best is to do good work for their clients. When they are not actually working for their clients they like to think about them, and what they might need, and talk to each other about them. None of the respective firms would have been in the room at this point if it were not for the fact that each believed that this deal would potentially deliver excellent results for clients. But rules are rules, and lawyers are scrupulous, and so the client angle of their discussions remained confined to what was already in the public domain – detailed enough to be tantalising and exciting, but inevitably a little generic.

Secondly, the team was at great pains to get this stage right – they were forensic in their investigation and articulation of the opportunity, because they wanted to know if there was a compelling story to be discovered.

And a third reason for the careful orchestration of these meetings was that the leadership team wanted to do everything they could to inculcate a culture of mutual respect, trust and collaboration right from the outset.

Three's a Crowd?

It is perhaps worth considering at this point the particular dynamic which resulted from the proposed merger being between three parties rather than two, and what prompted the firms to pursue this harder path. This approach undoubtedly added complexity and increased, apparently exponentially, the number of issues to be discussed and resolved. There were the inevitable challenges around aligning priorities and perspectives, and potential interdependencies and synergies which needed to be identified. There were different voices and histories to be understood. This approach also created uncertainty around perhaps the most fundamental point of all. In a law firm partnership, partners have the right to vote on whether or not to proceed with something as game-changing as a merger. This would in any event be a big deal for the leadership of a law firm, and in this case all parties were depending on a yes vote from all three firms – by no means a done deal.

Overall, though, the fact that there were three parties to the merger was a huge advantage – in fact, a critical factor in enabling the firms to do the deal that they wanted to do, the way they wanted to do it, and then to build the new firm that they wanted to build. Across the board, there was a consistent sense of orientation towards client relationships, a commercial approach and a truly distinctive focus on sectors and a depth of industry expertise. The clear theme emerging was that the firms were aligned in their purpose and could do business together. The very fact of conversations being three-way rather than bilateral shifts mindsets into being collaborative and drives towards the discovery of a middle ground, or a new and better way of doing things, rather than being binary and combative. There was no dominant party and no risk of any one culture overwhelming or subsuming another, and as a result, a better deal was designed and a better vision created for the new firm between the three than any two-way combination would have been able to achieve.

No Mud, No Lotus

Once the three firms had established that both the strategic and cultural fit was there, and that there was a compelling rationale for clients, then – and only then – they began to explore whether the proposed deal made sense financially. This process of establishing whether there is a deal to be done and a story to be created together is an unapologetically linear one (Fig. 6.1). The decision whether or not to ask their partners to do a deal of this nature, and on what terms, is of paramount importance for all parties, and exceptionally complicated. It is vital for clarity of thinking to keep the stages distinct and separate:

The financial due diligence, financial modelling and structuring phases of any transaction are clearly important *per se*, and they also helped to build trust – nudging the parties towards a deeper transparency and prompting conversations around, for example, their respective appetites for risk. The projected synergies as a result of the transaction also became part of the narrative for the new firm – but they were never the reason or the motivation for doing the deal.

Clients and Strategic Fit	• Does the proposed deal better enable us to serve clients? Are clients asking for what the new platform will deliver? Does the proposed deal align with the strategic vision of the firms and will it help us to achieve those visions, together? Do we understand how this deal will position us in the market? What are the risks and opportunities?
Cultural DD	• Do we like each other? Are we compatible? Do we share fundamental values? Do we see the world in the same way? Are our purposes aligned? Is our orientation towards the market similar? Can we imagine being in partnership with one another?
Commercial DD	• Are our businesses compatible and complementary? Where are the gaps? Where are the overlaps? Would this deal be value-accretive? What opportunities could we pursue together that we can't pursue alone? Do we risk cannibalising each other? What will clients think? What about conflicts?
Financial DD	• Are our businesses robust? Do we measure things in the same way – are we comparing apples with pears? What are we each bringing to the table and what is it worth? What are our financial attitudes and practices?
Financial modelling	• Does this deal, which we are now beginning to think we might like to do, make sense from a financial perspective? What are the opportunities and what are the risks? What choices to we want to make in the light of what we have learned so far about our cultures, and about the commercial opportunities?
Tax structuring	• If we are going to do a deal, what points do we need to think about from a tax perspective in terms of putting it together?

The CMS model for merger due diligence.

Figure 6.1 The CMS model for merger due diligence

What is more interesting for our purposes is how effectively and openly the three firms worked together in these areas to agree what could be achieved, and over what time period, bearing in mind the wider and more important considerations around being able to deliver a much-enhanced offering to clients, our go-to-market sector strategy and the commercial vision – and establishing a strong and attractive common culture for the new firm.

As discussed previously, the leaderships teams across the three firms actively wanted to plan, navigate and travel this journey together as part of the process of coming to a deeper understanding of one another – each other's stories, and culture, and values – and as the first step of their creation of something new together. While received wisdom might suggest that this was not the easiest route to take, in the context of the CMS merger, I saw it foster understanding and openness, and empower and enable the leadership to address some important cultural issues directly, early doors. In the words of Thich Nhat Hanh, 'No mud, no lotus.'

Or, in the words of Michael Rosen's children's classic, *We're Going on a Bear Hunt*:

Uh-oh! Mud. Thick, oozy mud. We can't go over it. We can't go under it. Oh no! We'll have to go through it . . .

The firms chose to walk through their whole adventure together, so that the story of having walked through the mud, as well as through the sunny uplands, would become part of their shared history – the cultural compost in which the new firm would grow.

Writing the Story

Once the firms had established that there was a deal to be done that all three parties wanted to do and that was compelling for clients, the team embarked on two major pieces of work in tandem – they began to pull together an Information Memorandum which would, in due course, make the case for the proposed deal to the partnerships in all three firms, and they began to consider and negotiate the terms on which the deal would be done. These two streams of work were very much two sides of the same coin – the 'story' of the deal had to inform the terms, but the terms of the deal would also become a critical part of the story.

It was agreed early on that the leadership would articulate one merger story in one single Information Memorandum, and make the case for the merger to partners in all three firms in identical terms. While this made the drafting more of a challenge – the three firms shared an overarching rationale, but each had a different perspective and emphasis that needed to be brought to bear – it was important that all partners considered and entered into a deal on the basis of the same information and with a shared understanding as to the rationale and the vision.

This was one of the fun parts. At the end of the day, City lawyers are deal junkies. There is the thrill of negotiation, and there is also the thrill of drafting, of laying down line after line, sentence after sentence, page upon page of clean, clear language. Of taking a concept and bringing it into sharp relief, pinning it to the page with the precision of Zorro, feeling the warmth of the freshly printed and sharp-stacked pages as they put on weight and bulk. So much industry! Night after night, weekend after weekend, the team pieced together the data and proof points, made the case for the merger and wrote it all down.

Meanwhile, the deal negotiation for the most part was straightforward, and always collaborative and amicable. Having agreed early on the philosophical principles which would underpin the financial deal, much of the time spent on negotiations was spent fine-tuning the application of these principles in practice. There were of course some more complex issues to navigate and it was also important to devote time to discussing governance structures and leadership roles in order to achieve fairness of representation, clarity of roles, and strong leaders in key positions.

I almost wish I could tell you that there was some heat. Conflict and tension are, after all, important refining fires in many relationships, and the key inflection points in the best stories. Back when I was a corporate lawyer, it was a rare deal that was done without someone banging the table, or walking out, and the eleventh-hour negotiations were often liberally sprinkled with elaborate military metaphors and swearing. But again, in this case, the direct approach and the unique three-way dynamic meant that the negotiations remained constructive and collaborative throughout. In the end, the final draft was produced and agreed in the small hours of a Sunday morning, after a weekend living on crisps and coffee in a building with no air-conditioning and which could only be accessed via the loading bay, because it was already being refitted to accommodate the new firm in anticipation of a positive outcome . . .

The Story Itself

This is the story that was told to partners in the Information Memorandum, about three sector-focused and client-led firms who, together, had a unique opportunity to create a new top City of London law firm.

It was called a new City 'powerhouse'. There was a photograph on the front cover of the document, taken by chance by a team member. She was facing east down the Thames on Southwark Bridge, just behind the Cannon Street building that would become the London headquarters for the new firm. It is sunrise, and the sky behind Tower Bridge is glowing, orange flames reflecting and flickering in the water like candles. Like hope and celebration. The windows in the Shard and the other buildings along both banks are golden, as though powered from within. Higher above, the clouds are gunmetal grey and menacing. It is a picture that is both powerful and beautiful. Firmly rooted in the City, grand and yet intimate. In bold white type on the bottom on the front cover, it says:

A New Top City Powerhouse

Differentiated by world class sector expertise

With outstanding global reach

Creating a distinctive, modern firm driven by technology

Underpinned by a 250 year heritage.

Look, already, how many stories the leadership team was starting to tell – shared history, journey, quest, ambition, perhaps even the threat of a shifting legal market lingering in those clouds . . . a picture is indeed worth a thousand words!

Inside the front cover, is a letter from the senior leadership of all three firms. In it, they tell the story of a new firm, uniquely positioned in London, throughout the UK, in Europe and internationally, ideally placed to provide clients with a distinctive offering driven by technology and sector expertise, underpinned by deep experience. They talk about the firms' compatibility in terms of culture and values, commitment to inclusion and respect, and about the power of our global platform, and then they lay out eight distinct threads of the vision:

- increased bench strength, scale and execution capability for all clients

- strong trusted advisor relationships and client focus – an ability and willingness to work end-to-end with clients through all stages of their lifecycle

- a dynamic, progressive and inclusive firm that embraces innovation and technology to transform client service

- differentiation in key sectors – bringing together some of the most respected experts in their fields and creating a deep pool of industry knowledge

- enhanced international offering for existing clients, and a platform for further international growth

- a centre of excellence for dispute resolution

- market-facing excitement and momentum, and empowered and inspired people

- immediate market leadership by virtue of the new firm's standing, scale, presence and expertise in core sectors.

Each thread is expanded on at length in the document, with evidence and examples. The hope and expectation was that every partner across all three firms would find at least one of those threads a compelling rationale for the proposed transaction. That something would resonate with their own narrative. But of course none of these compelling arguments could sell the merger alone; all are predicated upon an aligned leadership and a strong culture, and upon partners feeling that they could trust both.

The letter ends with the signatures of the senior and managing partners of all three legacy firms, bold and black on the white page. Leaders telling a story, and powerfully demonstrating their commitment to walk the talk.

The body of the document starts by introducing the main players. In any good story, there are compelling characters, bringing their own histories and personalities into the story. Neuroscience tells us that when it comes to our propensity to trust one another and make discretionary effort to help one another, the sharing of just two short personal interactions– *How was your weekend?* and *Isn't it rainy out?* for example – is sufficient to shift us from assuming threat to being reassured as to good intent. Albeit in a corporate vein, this part of the Information Memorandum sought to do much the same – to humanise the three firms, bring them to life, and to begin the immensely important task of building trust.

The document then expands at length on the case for merger – lawyers like logic and reasoned arguments, and so it was appropriate and important for the particular audience that the groundwork was laid in this way.

It then goes on to paint a picture of the vision for the new firm– always with clients right at the centre and, again, including plenty of concrete detail to meet the needs of the audience. The document talks about governance and management structures and who will be in leadership roles. It described the new firm in detail in terms of its geographic reach, and sector and practice group scope. It then describes the three parties' aspirations around the culture of the new firm and what it would be known for – exceptional quality, collaboration, innovation and technology, thought leadership, development and contribution to our local and global communities.

Finally the document sets out the terms of the deal and explains how risks have been assessed and managed. It concludes with details of how integration will be managed, and specific next steps.

Telling the Story

As the team were putting the finishing touches to the Information Memorandum – adding schedules, checking cross-references – and finalising, in the wee small hours, the last outstanding deal terms, the legal press was fizzing with speculation. Sixteen hours before the leadership team was scheduled to take the proposal to the partnerships of the three firms, the correct story broke in the press, at the end of a fevered afternoon.

But of course, long before negotiations concluded, and the story broke, the team had been working on internal and external communications plans. It was important to ensure that the deal rationale and strategy was communicated in a way that would emphasise above all the enhanced service offering for clients; the foundation upon which the merger was built.

The day before the story broke there was a briefing meeting held for all leaders. This took place in a London club in the traditional model (save with a more enlightened approach to women!), tucked away discreetly in the heart of the City. I remember the day as dark and low. The meeting convened around a horseshoe table in an upstairs room with dark wood floors, faded carpets and an assortment of upholstered chairs. It quickly became clear that there was a full turn-out, and so there was a bit of a family-wedding-type scramble for more chairs, and a passing round of cups of lukewarm, stewed coffee.

It is funny what ends up having that almost photographic quality in memory. For me, clearing my throat to address this higgledy-piggledy roomful of fifty people, all serious players at the top of their game, all nervous, some excited, some more contemplative – for me, that was one of the stand-out moments of the deal. It felt simultaneously that we had come so far – astonishingly far – and yet that we were only just beginning. The people around that table had centuries of stories and history between them. Many of them had never met before that day, and yet here they all were, about to write the next chapter together.

Notes

1. Krill, P.R., Johnson, R., and Albert, L., 'The Prevalence Of Substance Use And Other Mental Health Concerns Among American Attorneys', *Journal of Addiction Medicine*, 10: 46–52, 2016. Numerous studies in this area over the past twenty years prompted Martin Seligman, one of the leading thinkers in Positive Psychology, and Professor of Psychology at the University of Pennsylvania, to write a whole article entitled 'Why Lawyers Are Unhappy': Seligman, M.E.P., Verkuil, P.R. & Kang, T.H., 'Why Lawyers Are Unhappy', *Cardozo Law Review*, 23: 23–55, 2001.
2. Research suggests that analytical thinking can reduce our capacity to empathise and care, in part because our brains use different neural pathways for each and we literally have to switch between the two. Small, D.A., Loewenstein, G. & Slovic, P., 'Sympathy And Callousness: The Impact Of Deliberative Thought On Donations To Identifiable And Statistical Victims', *Organisational Behaviour And Human Decision Processes*, 103: 143–153, 2007; Jack, A.I., Dawson, A.J., Begancy K.L., Leckie, R.L., Barry, K.P., Cicca, A.H. & Snyder, A.Z., 'fMRI Reveals Reciprocal Inhibition Between Social And Physical Cognitive Domains', *Neuroimage* 66: 385–401, 2013.

Chapter 7

Starting Out

*Genius is 1 percent inspiration
and 99 percent perspiration.*

Thomas Edison

Getting to Yes

On a Friday morning in September, all three firms convened full partners' meetings to tell their partners about the proposed deal, share the Information Memorandum, and explain next steps. While the press coverage the day before had certainly taken any element of surprise out of the meetings, the atmosphere, in all three rooms, was nevertheless one of anticipation. The senior leadership of each firm presented the proposal to their own partners, in their own words, based on the vision and rationale as set out in the Information Memorandum. Partners were given all the documents and a schedule of many smaller sector, practice group and team meetings, and meetings with senior management which would take place over the coming weeks to allow for deeper analysis and discussion, after a weekend's reflection.

In order to ensure confidentiality, in the Information Memorandum and all the other deal documents, the parties were referred to by code names. Rather than decode the Information Memorandum for printing and circulation, the team had produced a covering letter with a secure peel-off squares, like your bank would use to send you your pin number. When partners left the meeting rooms after those first announcement meetings, the floors were littered with little squares of transparent paper, slightly static and dancing, like tickertape after a parade, or confetti after a wedding. A Big Day had come and gone.

Alongside those partners' meetings, a different meeting took place in a lower ground, low-ceilinged function room in a hotel near St Paul's, late in the day on that same Friday. This was a meeting for the heads of all business services functions across all three firms, to let them know about the proposed deal, although the basics were of course by now already widely known, and to ask for their commitment, hard work and discretionary effort to help the leadership team deliver it.

This was an early and very tangible example of what is often one of the more challenging aspects of deal-doing and subsequent integration.

In many major change projects, the people an organisation most depends upon on to help to shape the future story on the ground, and bring it to fruition, are the very people who do not know, and cannot know until much later, whether they will be part of that future. This is a profound challenge from a change perspective, and it seems clear that the Kotter model[1] of thinking about staff engagement only later, and about culture only at the end of a process, cannot work in such a scenario. The leadership team needs to help people to sense-make – to articulate their own stories and reasons for being there and to weave these into the story of the organisation.

In the context of the CMS merger, many of the most key members of the support teams were of very long standing with their respective firms and felt a great deal of passion for those firms and their history. Engaging with that part of them, and helping them to see their role in crafting a new future was, for some, motivation enough. Others saw the project itself – the chance to work on the biggest merger in UK law firm history – as a once-in-career learning and CV-enhancing experience. But others found the idea of the forthcoming changes difficult, and just being able to acknowledge that as part of the wider story was important – for the individuals concerned, but also for the wider firms as they saw the leadership team contending with their own becoming and trying to live out their values together. As we shall see throughout this story, the senior leadership took both the views and the wellbeing of everyone across the firm exceptionally seriously throughout the whole process – again, the central importance being accorded to the human aspects of merger. It really is impossible to overstate the importance of trust, respect and generosity and the difference they can make.

Meanwhile, practice group partners met together, management teams from each firm visited the other two firms to make presentations and answer questions. The leadership team ran wall-to-wall meetings, prepared screeds of Q&As and, as far as they could, kept press and speculation at bay until, late one evening, much sooner than anyone had expected, it was reported that the vote had gone through.

Telling Everyone Else

The team announced the formal approval of the deal to all staff and to the press first thing on a Monday morning a few short weeks later, and devoted the day to a series of townhall meetings explaining the proposal, the rationale, and its implications, and giving people a chance to ask questions and discuss. The leadership teams of the three firms were captured on film together, talking about why the merger made sense from their perspective and what they were looking forward to. This was sent to everyone.

The firms also had a targeted campaign to let clients know, via the best single point of contact across each of the three firms, about the proposed merger. Partners were really keen to be allowed to talk to clients, to identify opportunities and assess their reaction to what was being proposed. They had lots of compelling collateral ready to share about what the combined opportunity would mean for particular clients in particular sectors, combined credential lists . . . and, of course, there was a painstaking protocol to ensure that all necessary consents were obtained and all confidentiality remained completely intact.

Getting Organised

It is almost laughable now, but the strong desire and expectation of the team, myself included, was that it would be possible to announce the deal to everyone, talk excitedly about it for a couple of days and then ask everyone to put it out of their minds until after Christmas, while the deal team took a breath, reacquainted themselves with their families and spent some time with cold towels on their heads revisiting and finalising the plans for the period up to 1 May.

Of course, nothing of the kind happened. People were immensely excited. They had old friends in other firms. They had worked alongside or across the table from each other for years. And so they went for drinks, they exchanged notes, they made plans. Very quickly, it became clear that the team would have to be much more decentralised and laissez-faire in its approach than I, at least, had anticipated. Momentum and enthusiasm were ignited. To attempt to extinguish them would be both painful and foolish – it would be much, much harder to rekindle enthusiasm again later if it were extinguished now. We gave everyone the basic remind-ers and guidelines that were necessary in order to ensure that everyone behaved impeccably in terms of confidentiality and information sharing, and then left them to it. As time went on, we also allocated budget on a fairly ad hoc basis to support social events. The role of the central team became one of facilitating, advising and sharing best practice and good ideas across the business, while people embraced the merger story and made it their own. I was learning a lot, and fast, about being comfort-able with uncertainty and about being dynamic and evolutionary in our approach. This was a liberating and brilliant position to be in – not least because we had quite a lot of other things to think about . . .

Because many people also had questions – detailed ones that felt very pressing, lots of them, right away. Over the years I have learned – and the neuroscience clearly supports this – that it is important to respond to people's questions as far as one possibly can. Our brains want to be able to predict and make sense of the world. Even saying, 'We don't yet know the answer to that, but we are planning to discuss it on x date, and will aim to be in a position to share an answer on y date' helps people immensely to manage their uncertainty at a neurological level, and to become unblocked and able to act.[2] With the certainty that a question answered confers, peo-ple become active participants – able to create their own narratives and be part of the bigger story.

The team also had to learn, personally and institutionally, to deal with uncertainty and ambiguity. While creating, negotiating and deliv-ering a deal is intense and involves an immense amount of work, there is nevertheless a simplicity to it – one clear objective, a small number of inter-related workstreams to deliver in order to achieve the objective, a relatively small team with clearly defined roles. The matter of actually delivering a viable solution for a fully merged firm within six months is a different kettle of fish altogether, involving far more work, far more people and far more complexity. For weeks – literally weeks – I kept a notebook by my bed and would wake up several times a night and scribble down a note of something I had suddenly remembered we needed to consider. In time, and with the support of a team of brilliant project managers who had already developed a framework, we put a shape and structure around the tasks in hand and devised a series of workstreams, with project leads, subject matter experts and sponsors allocated to each, identified key deliverables, and milestones (see overleaf).

When I look at this workstream list now, all neatly nested inside itself like a set of matryoshka dolls (Fig. 7.1), I think on the one hand how gloriously organised the CMS team were, and yet on the other hand how it is no wonder they missed myriad little things – could they have spotted, for example, and incorporated into their estimates of printing volumes, the impact of the use by partners in one particular team of small, personal, desk-side printers? Should they have predicted that client footfall in Cannon Place would more than triple post-merger due to the sheer enthusiasm and curiosity of clients, dwarfing the aggregate of volumes experienced in all three legacy firms for the full year pre-merger?

There is an unimaginable amount of detail in every workstream. An unimaginable amount of thought and endeavour. Thousands upon thousands of small decisions, cumulatively driving towards a single goal. Delegation and a distributed approach to decision making become absolutely vital if everything is to be progressed and delivered. And yet everything has to be joined up, and hang together, and someone needs to have oversight, spot and resolve issues before they derail things, manage and connect the people who are managing and connecting other people who are managing and connecting . . .

clients
- shared clients
- sector penetration
- opportunities
- conflicts
- client liaison
- client data
- pricing and panel arrangements
- campaigns
- client relationship teams

risk and compliance

practice integration
- domestic
- international
- strategy and plans
- reporting
- relationships
- headcount model
- talent mapping

user experience

day 1

finance
- ops and systems
- reporting and accounting
- data flow

transitional services

secretarial model

document production

deeds and records management

vendors and procurement

premises
- Cannon place – including, eg client services, HSE, hard services, cleaning, security, mail, print, records, deeds, scanning, floor plans, catering,
- moves
- sublets

HR

IT
- infra and networks
- data, backup and DR
- applications
- Service Desk and support

A sample list of workstreams, similar to that used by the three firms.

Figure 7.1 A sample list of workstreams, similar to that used by the three firms.

... This perhaps explains why at this stage of the story, I felt that most of my own job was talking – I would go home daily sounding hoarse – and managing a frankly ludicrous volume of emails. The role of leaders in this season was very obviously storytelling, and story creating – to remind people what we were aiming for, and why and to help the operational decision-makers to proceed in a way which best fitted that aim, sometimes by directing, and sometimes by empowering.

Do It Yourself

Because the leadership team were determined, as noted above, that the process of creating the firm should itself become part of the firm's shared story, they were also keen that, as far as possible, responsibility for the operational execution and delivery of the merger and subsequent integration should rest with people within the firm. They judged the people who had been part of the legacy firms' pasts to be the people best placed to create the new firm's future. They wanted to give career-making opportunities to their best people.

This was the right decision. In practical terms it means that now, more than a year after the merger, the people who know every aspect of the new firm intimately are still right here. They know what the team did on merger, why they did it, what else they tried, what worked, what didn't, and so what the options might be next. That is incredibly important and practically helpful.

There is also a high level of ownership, responsibility and agency right across the business – a feeling that they created the new firm together. At a leadership level, all worked tirelessly to set the agenda, build confidence and deliver. At an operational level, every single person involved in implementing the transaction – in fitting out new floors of the building, creating marketing materials, working out seating plans, configuring software, negotiating insurance, building the press campaign – became part of the merger story. They brought ideas, skills and sheer hard graft, along with their own stories-so-far, and put them into the mix to create something new, and something that was different by virtue of their very input and involvement.

The other advantage of having had home-grown teams hands-on throughout is that they got to know one another across legacy firms and functions. This was a fierce and fast kind of knowing – people under pressure, unsure of their own futures, working together for the first time – they were up close and personal for good and for ill and so by the time we got to 1 May and the three firms officially embarked on their future together as a single firm, there were large groups of people who had been working together for a long time and knew each other well.

This approach, though, is not without its challenges. In particular, it is important to ensure that individuals and operational teams who are taking on additional responsibilities in this way are supported, emotionally and practically, in managing their own bandwidth and energy. It may, for example, be necessary to be very explicit from the outset about which parts of 'day jobs' can and should be set aside for the duration of the implementation project. It may be appropriate in some instances to provide back-fill support. In a flat hierarchy populated by people with high levels of responsibility and intrinsic motivation, it may not be immediately obvious how busy someone is – people may keep their heads down and keep going rather than complain or ask for help from other busy people – and so it is incumbent on leaders to be proactive in this area, and also to be prepared to take a moment to breathe and reflect. Sometimes there can be a sense that by the time the team stops to address anything, things will have moved on in any case, but this can perpetuate any problems and can come at a cost to the wellbeing of an individual and the wider team. Good leadership sometimes means pressing pause.

It is also important to prepare the operational teams for the likely reality of the first few weeks after any major change project 'goes live'. In a sense, any preparation phase, however intense, is 'the easy bit' inasmuch as there tends to be a higher level of control, a detailed plan, fewer variables, a close and committed team with a high level of shared expertise, and the buoyant mood that comes from anticipating a good outcome. (We were discussing around the family dinner table recently 'favourite feelings'. One of my daughters volunteered 'looking forward to something' as the best feeling in the world. I can relate to that.) In my experience, though, those first weeks after go live are inevitably difficult, and all the more so for tired teams who feel an acute sense of ownership and responsibility.

In the present case, while it is certainly true that some of those on the CMS teams at times had challenging workloads, and that the early weeks post-merger were challenging, it is also true that the sense of teamwork was so deeply rooted, and the commitment to deliver excellence every time so unwavering, that the team continued to dig deep and to build strong solutions over the weeks that followed. It was remarkable to witness. As solutions emerged, confidence in turn grew and the momentum for success built. Again, 'no mud, no lotus'.

Do It Once, Do It Fast, Do It Right

The only reason for time is so that everything doesn't happen at once.

attr. Ray Cummings

In the specific case of the CMS merger, part of the challenge for the operational teams arose by virtue of the sheer scale of the firms' ambition, and the short timeframes they committed to. So, why go so fast? Why was the leadership so ambitious in the scale of integration they sought to achieve within a six-month timeframe? Plenty of firms have taken a slower, more incremental approach – operating across multiple locations, on different finance and CRM systems, with different terms for employees, for a year or eighteen months post-merger – and that where there were only two parties! That was never the vision in this case. For the biggest merger in UK law firm history – and a three-way one at that – the leadership were determined to deliver a co-located, single law firm, operating as one team, on one set of systems, and delivering a fully integrated service to clients, right from Day 1.

The main reason for this was because they believed that their story depended on it. They were aiming to be a future-facing law firm, distinguished by the depth of their sector understanding, their closeness to clients, their capacity to problem-solve and get things done. They could only deliver on that if everyone was truly working together – physically side by side, incentivised together, on the same project for a shared client.

The leadership also understood that the real work – and the real value– lay in bringing the firms together at the human level – building a culture together, learning from each other. And they believed they stood the best chance of doing that once they had addressed 'everything else' – by which they meant systems integration, alignment of terms, co-location and general operational snagging. There was an acceptance that the sheer process of merger, however well planned-for and managed – and this one was – would inevitably bring x amount of discomfort, and that the better course was going to be take it on the chin and work hard at solutions in the good wind of this transient season, rather than string it out over a protracted period and watch goodwill and motivation slowly drift away. This would likely mean a challenging few months, but by six months in, the new firm wanted success to be theirs to play for, no excuses.[3]

Shaping the New Firm

Throughout the pre-merger period, there was a huge amount to be done, with a very clear back-stop deadline. Working in this sort of context can easily lead to a heads-down, task-focused mentality but, again, it was vital that everything was joined up, and that means speaking to each other. It was also vital that close working relationships and trust between the management teams of the three legacy firms should be maintained and deepened. To help with both of these things, the leadership put in place a light-touch management and governance structure specifically for this period.

An Integration Leadership Team, comprising the senior and managing partners of the legacy firms, had overall responsibility under the terms of the merger agreement for key strategic and client-related matter matters. This team was absolutely central and became all the more so – these leaders set and reinforced the firm's strategic direction and became the figureheads, the bellwethers. They played a key role in helping to build an understanding of the legacy firms' respective histories and cultures and how those might play out, and in building trust throughout the business. They became the keystone in the new firm's Board, and advisors to the newly appointed leaders of practice groups and sector groups – able to give the broader context and support decision making.

There was also a Core Integration Management Team, which considered and made recommendations on the myriad operational decisions that had to be taken in short order during this period. They advised on, for example, client consenting processes, team structures, project management, standards and precedents, premises, Day 1 protocols, communications strategy, TUPE and benefits, email addresses, leaderships workshops, benefits, brand . . .

A common bear trap for groups of this kind, who tend to be asked to troubleshoot and to get into the weeds a bit, is that it can become easy to lose perspective and the conversation can start to feel a little negative. Alive to this risk, with the team at CMS we instigated a standing agenda item of good news at the top of the agenda. It sounds silly, but I cannot recommend this approach highly enough. It is a tiny thing, but it shifts the focus significantly and cultivates a posture of appreciation and perspective which, at CMS, proved to be foundational to the way the new firm does business. The operational teams also maintained a long email chain simply called 'Progress' to which everyone would try to add something every evening, just listing perhaps two or three ways in which they had made progress that day, however small. These were the single steps on the journey, and marking them was important and perpetuated gratitude, motivation and momentum.

Perhaps inevitably, the hardest aspect of the period post-announcement was the need to restructure the business in advance of 1 May. Hardest by far, of course, for those directly impacted by the uncertainty and then by the outcome – and hard for those who were tasked with keeping the show on the road day-to-day and planning a positive future for the firm amidst a restructuring process with all the uncertainty and emotion that brings. Here was a true litmus test of the culture the leadership team was committed to building – it was vital that all involved should approach these changes with the utmost respect, trust and generosity. The leadership team was committed to ensuring that everyone received the practical and emotional support they needed to help them to conceive of, and then transition to, the next part of their story.

Project Triangle

It was also vital that, in the period from October through to May, those who were going to be part of the new firm's future should feel engaged, excited and motivated . . . while also maintaining business as usual as far as possible across all three firms and enabling people to remain focused on clients. This is never an easy balance to strike and it is hard to help people to direct their attention and focus in the best way. Apart from anything else, it is often necessary to throw a ton of information at people. All of this requires very purposeful leadership, hands-on management and exceptional communication skills.

For CMS this meant that, in due course, they had to supplement the happy grassroots opportunistic approach that was prevailing with a more structured approach. Something which almost always helps in my experience is to make all the work and activity that is taking place over this period 'a thing', and this is what the team at CMS did. They called the thing Project Triangle, for fairly obvious reasons, and designed a logo which had three differently coloured triangles converging around a common apex. They developed a tagline which encapsulated the spirit in which they wanted everyone to approach the next few months, and which also provided a framework for communications (Fig. 7.2):

INTEGRATING FOR EXCELLENCE
OPEN MINDS /// OPEN DOORS /// OPEN OPPORTUNITY

Figure 7.2

In a single day in a white-walled room, a fantastic communications team drawn from across the three firms wrote down on neon post-it notes everything they could think of that they might want or need to communicate over the coming months. They grouped the hundreds of ideas into themes. They talked about how to manage the communications in 'seasons' so that people were not inundated and, more importantly, could see themes unfolding which would help them in developing their own stories.

The team developed a microsite which could be accessed from all three firms, and a weekly newsletter which would push content out to everyone. These were two of the primary means by which the leadership team could tell the story of the merger process in real time and give people a sense of the different seasons that they would be travelling through. Every edition of the newsletter started with an overview from the senior leadership team. This gave everyone a strong sense of the voice of the future firm, and provided the perfect platform for positioning any key messages.

The first edition introduced the concept of Project Triangle and the strapline, and thereafter the newsletter was structured to emphasise those three elements:

Open Minds – telling people some new information or asking them to do something

Open Doors – giving people an opportunity to participate in something, often social

Open Opportunities – showcasing client successes, and seeking input on joint initiatives with clients.

The team also launched a merger contribution award, which was awarded weekly to individuals or teams who went the extra mile to progress integration or who made a suggestion that was adopted. Award-winners and their ideas and contributions were showcased in the newsletter every week.

Tell Us Something We Don't Know

Early in the new year, a few months ahead of 1 May, the leadership team conducted a cultural diagnostic survey. The primary aim in doing this was not so much to learn something new, although there were of course some interesting insights, but to equip the leaders with a neutral and common language for exploring some key themes, and thus to aid integration.

Much had already been made, and rightly so, of the three legacy firms' apparent 'cultural fit' with one another. Fit, of course, does not mean homogeneity, and nor did the firms want it to. What was meant when the leadership talked about the great cultural fit was, first, a similar approach to market across the three firms – client-focused, deep in industry sectors, commercially savvy – and, secondly, a similar energy level and degree of informality. It was this second factor in particular that led to the difficulty in silencing dinner conversation right from the outset, lots of laughter in introductory meetings, and a general sense that we could do business together.

The diagnostic tool went deeper than this, though, into the less-spoken-of things about how power is concentrated or dispersed, how information flows, how decisions are made, what is celebrated. In my experience, this sort of insight can be really helpful, not so much because any of it is directly actionable, but because it can help teams to understand potential sticking points and to navigate issues. What's more, the very fact of conducting such a survey emphasises the central importance of culture and, in the present case, really helped to reinforce the message that the leadership team were deeply respectful of the cultures of all three legacy firms and were committed to seeking common ground to stand on.

In the same spirit, in the run-up to 1 May, the leadership team ran a series of workshops in which they asked people to explain what they most valued in their current firms and would wish to preserve, and what they most hoped for, and feared, in the new firm. In these sessions, people were also encouraged to talk about change, what to expect, how ready they felt, and what more they, and the leadership, could do in preparation.

The workshops themselves were an important part of the process and the story – they enabled every person in the firm to feel that they had had an opportunity to participate, to make their voice heard and to say what mattered to them. But the real value was in the sheer number and richness of the stories, anecdotes and insights into history that were shared. They gave a much deeper insight into the three firms, much faster, than the cultural diagnostic or anything else. There was something profound – almost sacred – about some of those workshops. People were generous and real and vulnerable. A hushed intimacy was created when people described, so vividly, the tiniest incident that had been a defining moment, summing up what their firm meant to them. An act of kindness, a moment of triumph, a silly tradition. It felt as though people were moving towards first of May with a brightly woven tapestry of stories streaming out behind them, and a fistful of richly coloured threads in their hands.

Leaders were encouraged to arrange a wide variety of social events in a wide variety of configurations in the run-up to 1 May – one-to-one coffees between partners, drinks for trainees, team quizzes, lunch and learns . . . There was a long list of all the ideal connections that should be made, but in truth the precise order and manner in which things unfolded was less important – the most important thing by far was to begin building as many social interactions as possible, as deeply as possible – in order to also start building trust across teams and to create psychological safety for everyone in advance of all the changes that were still to come.

Getting to Know You

Practice group leaders, team leaders and sector leaders had spent some time together in the run-up to vote, both socially and in order to identify and articulate the nature of the opportunity in their area. But there was still a huge amount to do in order to really figure out the proposition in each area of the business, and to begin building relationships in the leadership teams.

Now. Law firm leaders are invariably time-poor and in my experience much prefer to focus on their clients and their legal work, rather than devoting their time to business planning or to developing their 'leadership skills' unless they are persuaded that there will be very clear positive outcomes from doing so. For now, the challenge was to 'do enough' planning in a really focused and productive way to ensure a flying start on 1 May, and to get the leadership teams sufficiently well bonded that they were able to function well as a team from the outset.

The senior team engaged a consultant with particular expertise in leadership and neuroscience to design a programme around her new model of neuroscience-based effective team leadership, and then deliver it to each of the leadership teams. It proved extremely useful in meeting the teams where they were at, allowing them to focus on the areas that were most relevant or pressing for them while also opening up their perspective to consider other angles and areas. In most of the sessions, leaders ended up talking about identity, relationships and clarity, and the role of the leader in curating and creating these for a team. They talked a lot about storytelling. These common themes also meant, again, that they had a common language and way of thinking about the world that they could refer back to, and this was very useful.

Alongside this, leaders who were co-leading groups or teams, did some further work using a strengths-based approach to help them begin thinking through how best to work together to lead their teams.

In relationship and leadership terms, this work gave the new firm a good foundation. It also yielded enough substance in terms of business priorities and capabilities that teams were confident that they could put in place good, robust business plans in advance of 1 May.

Information Overload

As 1 May drew closer, the amount of information that it became necessary to share, and the number of actions that the teams actively needed people to take, increased exponentially. This is inevitable and, I think, unavoidable, in any major change programme. People at this point begin to struggle with information overload. On the one hand, there remains uncertainty and people want to know more. But on the other hand, the sheer volume is such that it becomes difficult to unearth the most salient pieces of information – things in respect of which one is required to act, for example, or to express an opinion, or exercise choice.

The team at CMS set up a new 'Integration in Action' email address which sent emails in scarlet block capitals when they really needed people to do something. They pressed leaders to cascade information informally, in team meetings. They kept the newsletters as short and punchy as possible. Nevertheless, despite a great team, a thorough plan and great execution – undoubtedly the best I have seen – what they had on their hands by this point was essentially a cacophony. In a sense, though, this was the most positive of signs – the firms were getting to the point where people's own stories, their practice and team specific stories, the various workstreams and threads of the merger stories, and a whole lot of extraneous details were coming together . . . indicating that they were ready for the next act.

Notes

1. Kotter, *Leading Change*, HBS Press, 1996.
2. See for example Scarlett, Hilary, *Neuroscience for Organizational Change: An Evidence-Based Practical Guide to Managing Change*, Kogan Page, 2016.
3. There was also a financial imperative around both bearing merger costs and realising the resulting savings as promptly as possible. A survey conducted by PwC in 2008 found that 82% of respondents reported favourable cash flow results from faster-than-normal integration.

Chapter 8

Making It Happen

Don't aim at success. The more you aim at it and make it a target, the more you are going to miss it. For success, like happiness, cannot be pursued; it must ensue, and it only does so as the unintended side effect of one's dedication to a cause greater than oneself.

Viktor Frankl[1]

Le Weekend

The weekend of the big move was the most anticipated and planned-for of the year amongst the operational teams. They had taken to calling it 'Le Weekend' – there was an almost festive atmosphere surrounding it. It was the May Bank Holiday weekend, and they had three days to change the world.

The last Saturday in April was the sort of high spring day where everything seems to warm and loosen and lean forward into summer. By mid-morning there were upwards of 150 people in the Cannon Place office. The post room team, the senior leadership, various partners' families. Everyone had hoodies, orange or blue, with the project logo on the front. There was a Le Weekend playlist on Spotify, and by the end of the weekend some had whole flashmob dance routines worked out. These tiny, ostensibly silly, things really encouraged a sense of team and occasion – they are important details in the story. There were packing crates, IT kit, chairs, boxes everywhere. A mission-control room full of flipcharts, post-it notes and pizza boxes. A whole meeting room dedicated to name badges – one for every person in the new firm, to be put on their new desk, once the new desk was built. Floorplans, easels, goodie bags, conference phones.

So began an absolutely spectacular team effort – a final sprint after months and months of planning and groundwork to deliver what, with hindsight, was a truly staggering outcome – a fit for purpose building and systems, operational and tested, to enable the new firm to hit the ground running on Day 1. I have scoured the market and literature since and have found no equivalent example that comes even close to achieving what the team at CMS achieved in such a short timeframe.

The team hit snags of course. For example, there was a problem with chairs. Chairs were the socks of the merger weekend. A team took all the chairs from the other buildings and brought them to Cannon Place. Half of them disappeared en route. The team took them in the goods lift from the second to the sixth floor. Another half vanished. It was all very odd.

They kept going.[2] Later, there was apparent shortage of extension cables. Someone went to every Maplin in London and bought the lot.

Those wearing fitbits were astonished and quietly smug to find they had clocked up the length of a marathon, indoors, in the course of the weekend. There was lots of laughter, shoulders slapped, tears, hugs, naps. The lead project manager, right in the middle of the web – buzzing, calm, all eyes on him. And all of it recorded by a reportage photographer, capturing for posterity the immense human effort, commitment and, yes, love that was poured into the creation of this place we call 'the office', this Fata Morgana we call 'work'. It reminded me of a barn raising. That scene in *Witness* with the soaring violins, and the framework of the barn with the men standing on it, up against a clear blue sky.[3]

Day 1

The first day of business for the new firm was 2 May. It was a bright spring morning across the whole of the UK – Scotland, Manchester . . . the new signage on the front of the Sheffield office glinting silver in the sunshine . . . Bristol, Reading . . . London dressed to impress in buttery light and high, tousled clouds. Just another day like any other, but also The Big Day.

The Day 1 team was in early, following the detailed plan to the letter, and wearing lime-green polo shirts, to make them instantly identifiable as 'people who could help'. I wore mine over a new dress and with sparkly silver sandals, which really were a bit too much for the office, but matched my mood and seemed perfectly suited to the sparkle which seemed to pervade everything. The London office was immaculate. There was not a hint of the absolute upheaval which had prevailed only a few hours earlier – in fact, there was that extra stillness that sometimes descends after a party, as if everyone has hurriedly scarpered, sweeping the place clean as they went. You half expect to open a cupboard to grab a tea-towel a few hours later and have two sleeping partygoers tumble out.

Every desk had a chair, a docking station, a Surface Pro which had been logged into and tested. On every chair, a goodie bag; on every goodie bag, a name badge pinned; in every kitchen, towers of new mugs. Easels holding floor plans in the lift lobbies of every floor; little blue 'you are here' stickers placed at thirty-two different 'right spots'. All of this new-ness inciting the same mixture of excitement and terror as a brand-new schoolbag sitting in the hall, the night before the first day of term.

At 8am, I went and stood at the top of the escalators at the main entrance, holding a spreadsheet of names. I couldn't believe my cheek and luck in having gatecrashed the team who had hands-down the best job of the day.

'Good morning,' we said to everyone, 'Welcome to CMS. Do you need help with where you are going?'

A slightly oddly-phrased question. We had thought that 'Do you know where you are going?' sounded too accusatory, and 'Do you need help with finding your desk?' somehow sounded that we were ramming eve-ryone in like battery hens. 'Do you need help with where you are going?', though, made me smile.

Doing this job meant that we were able to see absolutely everyone as they walked in, to assess the mood, to ensure a warm welcome. It also gave me something practical and useful to do, when I would otherwise have been pacing the floor, getting in the way. My hands were shaking as I checked through the spreadsheet to let people know their desk locations. I felt, tot-tering there in my daft heels and green shirt at the top of a steep escalator, as though I was balancing on the fulcrum between everything that had happened so far, and everything that was about to happen. I'm not sure I have ever been so simultaneously happy and daunted in my entire career.

The day went as well as it possibly could. Many people were at their desks and up and running well before mid-morning. 'I'm sending you this email to show you I can send emails!' . . . went more than one email that crossed my desk.

The senior leadership team had sent a powerful email to all partners immediately before the bank holiday weekend, reminding them of their vital role as leaders in the new firm, empowering them to make it work, asking them to be visible, positive and available, and giving them some practical tips on how to help their teams manage the enormity of the change. Partners were everywhere and all over it – upbeat, introducing people to one another, holding impromptu huddles around desks, cracking on with their work.

The team spoke on the hour with every other office just to check in, enjoying the end-of-day-one round-ups from Asia and then the Middle East as further encouragement as the day progressed. All was well. There was an almost holiday atmosphere. At 4.30pm, there was champagne and cake in the London office, and the premiere of the Day 1 video. It has been played so many times since that I almost expect it to have the crackle and jumpiness of a wartime newsreel. It features people from across the new firm, talking seamlessly together to camera about the future and what it looks like.

'One firm. Going Places. Creating Something Special.' it ends.

The new firm was indeed going places – great places. But I don't think anyone fully appreciated at the end of that gentle first day just how fast and bumpy the road to greatness was about to get.

Reality Bites – The First Fortnight

In every major change project I have worked on, there is a moment of 'go live' euphoria, just like the new firm's glorious Day 1. In every case, as in love, as in all of life, that initial euphoria then wanes, and is replaced by a period with a different kind of energy. At CMS, the very first few weeks of the new firm were permeated by an atmosphere which was still electric – excited and positive and full of opportunities – and yet also somehow arrhythmic or aflutter – everyone was so committed to getting everything right, and the stakes felt high.

The leadership team at CMS were deeply cognisant of and respectful of this period and steered the firm through its complexity with a marked care and confidence. What is perhaps most notable with the perfect 20-20 vision of hindsight is how, despite the unfamiliarity of their new context and some of the inevitable snags that were emerging around them, the lawyers managed to maintain exquisite, upbeat focus on delivering service to their clients with, in almost every case, no perceptible wobble. This is a huge testament to the focus and clarity of purpose of the firm's people, and testament also to the confidence and calm that the leadership conveyed.

In practical terms, there were – inevitably – some 'snags'. Some things were annoying for users, but easy to fix. Some of the internal meeting rooms still didn't have conference phones in them, for example. That was easy. A team wheeled around trolleys loaded with phones, followed a six-stage set-up process, and within an afternoon had 'solved' that problem.

But the most pressing issue was harder to solve. The biggest challenge we had by far was in relation to IT. Again, this has been true of every major change project I have worked on. To be clear, in this case, what the team at CMS achieved in relation to the integration of three firms' IT systems is truly mind-blowing. The sheer scale of what they were doing was enormous and required vast volumes of data to be migrated between servers. This was a big ask even in test conditions, but in a live environment when the systems were also bearing the load of people doing their day job, emails being diverted to new addresses, people looking online for help with new applications and systems, things snarled up and slowed down. This in turn led people to call or email the support teams asking for help, and soon these systems too became strained. It is easy to see with hindsight how things could snowball from a systems perspective – one system tripping another system into overload.

That was what happened from a systems perspective. From a human perspective, it was difficult. Generalising horribly, lawyers are motivated by doing a good job for their clients. They have exacting standards for themselves and others, and their overarching goal is to deliver the service their client needs. From their perspective, it doesn't matter a jot if 99% of what needs to happen in the IT systems has happened perfectly, if the 1% that is not working might delay that urgent email to that important client. It is hard for them to accept anything short of perfect delivery, and this on top of the stress that comes already from being in an unfamiliar environment, with new people and no established network or shortcuts.

Now consider the perspective of the support teams who were working on solving the underlying problems. Many of them were exactly the same people who had been working flat out in the months and weeks prior to Day 1 and had been part of the Le Weekend team. They, too, took immense pride in their work and were focused on getting things working. They were proud of being 99% there, and deeply frustrated about the 1%. They wanted to get on and fix it.

Bottlenecks and Bloody Marys

One of the things I loved about my role with CMS was that I was privileged to inhabit two worlds. As a former lawyer, I understand intimately how lawyers think and what is demanded of them in the environment they work in. I love their focus on quality, their intelligence and rigour, their ambition to deliver brilliantly for clients every time.

But when I was a kid and involved in a youth theatre group that put on shows on the Edinburgh Fringe, I was always happiest in a black t-shirt, organising props and costume changes in the wings, listening in on the sound deck, or perched on the lighting rig working the spotlight. The same has proven to be true in my career. I love being on the business side of the business, helping to make it happen. I like that I know my way around the back end of the finance system, and chose the colour of the paint on the walls.

As a result, during this early post-merger period, I felt immense empathy for everyone involved in every aspect of every challenge we found ourselves facing in those early weeks. And because I had also been intimately involved in many of the how/what/when decisions which to some extent had brought us to this place, and was now in possession of more operational detail across more areas than many in the organisation, I also felt immense responsibility.

Up to a point, this was a completely healthy and actively desirable frame of mind for a change leader to be in – big change projects *need* empathic people who take responsibility! But on the CMS project, I fell for a short while into the trap that I have seen countless others fall into and positioned myself essentially as the gatekeeper between the two halves of the organisation on operational matters. I wanted to protect people from one another: tired business people from frustrated lawyers; frustrated lawyers from tired business people who did not yet have solutions

'Talk to me,' I basically said. 'Talk to me, and I will get it sorted.'

This of course worked extremely well for the handful of people I was able to help personally, but it was hardly a scalable or sustainable solution.

Everyone on the team had to work out how best to use our time and resources to help as many people as possible make a success of these early weeks, and it fast became clear to me that I needed a different mindset. To stay focused on the past rather than the future – *Should I have predicted this? Should I have done more to prepare people?* – kept me reactive and defensive. It also betrayed a wayward variety of responsibility that risked being too controlling, and so potentially disempowering for people around me. I did not trust people's resilience enough. I had forgotten that the mud of discomfort is a necessary part of everybody's own change story. I was trying to make everything okay right now instead of letting people work their way into a better future.

Of course, it is hard for anyone to be in the fray during an intense period of change and not feel a little bruised. In Chapter 15 on Energy, we explore strategies for building individual resilience. I, too, perhaps suffer from that lawyer's tendency towards perfectionism, and at this point, I had allowed some of the basics to slip – sleep eluded me. I was powered by coffee, adrenalin, and handfuls of jellybeans from the meeting rooms . . .

On a Wednesday morning around a month in, I had a gap between meetings and I needed a breather. I hit the Thames Path and walked west, the river quiet and sleepy silver, the air already warm. I crossed the river and ducked into a hotel that serves a mean breakfast, complete with a Bloody Mary. I sat in the window in the sunshine. The vodka and spicy tomato brought me up short like a slap, the strong black coffee was like fortification, the salty, hot fat of the bacon, the sweetest sort of comfort. My sanity and years of experience fixed me with a stern stare across the table: 'You need to get a grip,' I said to myself, 'it was never going to be perfect on Day 1.'

Getting a Grip

That Wednesday morning was a mini-turning point for me. The facts had not changed, but I got back that scintilla of time between thought and action that is the aim of so much mindfulness practice. In that narrowest silver sliver, I rediscovered a degree of detachment, the capacity to prioritise, the power of choices and my sense of humour.

I learned a huge amount by observing the behaviour of the senior leadership team and its impact on people across the firm. Around this time, amongst the core team there emerged a very fluid and informal sort of daily, sometimes hourly, priority setting, based simply on asking 'What's best next?' and then doing that thing.

The senior leaders, each in their own way, clearly brought their entire selves to bear to tackle the challenges we were facing. Looking back, this obvious, authentic, humanity was a key part of the merger's ultimate success: yes, there were some challenges, but nobody could doubt that the leadership cared, and would show up time and again and get their hands dirty to fix things. This also served to ignite bravery and resourcefulness in others. There was a certainty, a sureness of purpose, a confidence, sometimes an almost childlike glee and curiosity that was compelling and infectious.

How Do I Love Thee? Let Me Count the Ways

The list is the origin of culture . . . What does culture want? To make infinity comprehensible. It also wants to create order.

Umberto Eco[4]

The weeks and months after 1 May were prime mating season for snagging lists. Lists begat lists – lists of the main lists, lists of the things that had been dealt with from the lists, that would be dealt with as top priority, that would not be dealt with. Lists of how people felt about how things on the last list had been handled.

The items on the lists varied from the serious – 'My computer doesn't work at all' – to the ostensibly less so – 'There's no avocado at breakfast'. The team worked through everything, fixed what they could immediately, put harder or longer-term things on another list, and in any event tried always to go back with an answer.

A group of partners from across the practice was convened to be the Systems and Process Improvement Team.[5] This group tackled some of the more material and urgent issues, hard and fast – conflicts checking, new matter opening, billing processes, front of house, CRM – and their very existence proved to be extremely useful; again, both in signalling the seriousness of the commitment to getting things right, and because this was basically more hands on deck, and a group who were motivated and committed to getting these issues sorted.

After a while, though, it was time to call time on lists. Lawyers love a list more than most, and are generally wedded to the idea that there is a right answer to everything. The ambition inherent in that thinking had brought the new firm a long way already – but now people risked becoming too inward-focused and too negative in their thinking. It was important to carry on striving to address issues, but people also needed to get their heads up and move on.

The First 150 Days – Making a Plan

And so, around six weeks into that first period, the leadership team drew breath sufficiently to begin turning their attention from react-react-react to articulating the plan for the short to medium term. The team had of course already done all of the groundwork on this prior to merger – they knew where they needed to get to, and by when. But it was important to refresh the plan in the light of those early-week experiences, to ensure its relevance and refine priorities. The team also had to make a judgment as to when and how to begin talking about the future in order to inspire people and ease them forward, rather than simply creating a backlash about the ongoing difficulties in the present.

The team articulated a plan that was as straightforward, clear and direct as possible – the aim being was to reassure, engage and keep moving. It was as simple as ABCD – they would focus on continuing to address the ongoing issues and help people across the firm to build relationships. The firm would increasingly turn its attention back out towards clients, and the leadership team would ensure that everyone was equipped to represent and deliver the new firm. Finally, the leadership team would begin to develop the new ideas and projects that people were advocating for for the future. Six actions under each of the four headings; twenty-four in total. Named people responsible for each one. Weekly reporting on progress against each action, shared with the whole firm, and a detailed monthly report to the Board. The plan was delivered in hard copy to everyone's desk, and stuck up in tea points[6] on Day 50, with 100 still to go to deliver the plan . . .

For the most part, this approach worked very well. The leadership team timed the distribution of the plan so that it immediately followed the global partners' meeting which had given partners an opportunity to lift their heads out of the daily grind, to remember again the scale and ambition of what they had signed up to, and to see the bigger – and still very compelling – picture. The leadership time also did a good job of listening, and reflected back what seem to be the 'right' four themes in terms of areas of focus – they resonated with people and were easy to remember. The plan became a really useful way of talking to people about next steps, priorities and their own issues.

Paper, Stone, Scissors – The Best of Three

One of the mantras that was maintained throughout the entire merger process reflected one of the senior leadership team's guiding tenets – namely that in creating the new firm, they would take 'the best of the three'. By this they meant that there was no dominant party, and no monopoly on good ideas. They wanted to take the best ideas from everywhere and incorporate them into the story of the new firm, and to be respectful of, and preserve, the best of the three firms' histories and legacies.

During the early weeks, despite this guiding tenet, the operational teams had had to make a number of pragmatic decisions based on what was immediately possible or most expedient, and not necessarily 'best'. The clear commitment in the 150 Day Plan to begin developing new thinking addressed this head-on, as did the establishment of the SPIT team. With hindsight, though, I wonder whether the leadership team might have spoken, instead, about a new 'fourth way', that would include, but also transcend, all of the best thinking and practice from the three legacy firms. A 'fourth way' is clearly less pleasing in the context of triangles, but the new firm that was being created was fundamentally different, its ambition was great, and the industry around it was changing fast. Increasingly, therefore, the leadership team found itself devoting its collective energy and imagination to creating entirely new ways of doing things, rather than trying to impose old thinking or ways of doing things on the new firm. 'Best of Three' took the firm to the threshold of its merger, but it was 'better than any of the three' that would drive it forward.

Keep Talking . . .

Throughout the summer, the new firm created lots of opportunities for people to learn about the firm and, more importantly, to get to know one another. The team deliberately tried lots of different approaches to appeal to as many people as possible – pub quizzes, 'speed-dating' networking sessions, breakfast seminars and volunteering opportunities. A 'suggestions' email box. Sounding-board groups and think tanks. Practice group socials were encouraged, and 'huddles' were invented – groups of people at neighbouring desks who could have impromptu lunches and coffees. There were coffee lotteries in which people could be randomly assigned to meet someone else from a different part of the business for coffee and a half-hour chat. The firm funded new networks, choirs and sports clubs. The senior leadership team themselves met groups of partners from across the business three times weekly. There were daily emails introducing partners to everyone across the firm. An immersive opera workshop session was held to teach leadership and impact to junior partners. Some things worked brilliantly; some things worked less well. The teams did their best and then trusted that anything to get people talking was ultimately for the good.

At the half year, once the leadership team had purposefully allowed things evolve and take their own course for a bit, they kicked off a series of twenty 'Open Door 2.0' sessions across the business. These were the first semi-formal 'check-ins' across the business as a whole since the merger, and were undertaken for a number of reasons. There was a desire to honour the commitment given in the first, pre-merger, Open Door sessions and give people a chance to share their reflections on their journey so far. The leadership team was also keen to begin putting some words around that inchoate thing called culture, in order to give some structure to the firm's story, and a framework within which people could tell their own stories. They also wanted to begin excavating and articulating the firm's purpose, so that they could root the work that was under way on the firm's business strategy and on the new proposition for our people firmly in that.

The stories, reflections and ideas that the team got from those sessions were incredibly insightful, potent and useful. People are amazing. Between them, a random selection of a dozen people in a room, from across the business, generally know the answers. The job of leaders in such fertile circumstances is simply to harness and direct the vision and daring of the people around them.

. . . Whoa, Keep Talking

In telling you some of the stories of the CMS merger, I have talked almost exclusively about internal matters. That is because that was my story – it was my job to focus internally so that the senior leadership team and all lawyers could be freed up to focus as much as possible on clients, and on ensuring that the benefits of this merger for them were being clearly communicated and abundantly delivered.

It is really important to understand, though, in order to understand and contextualise the internal story, that from Day 1 and right through the first year, the firm enjoyed a huge amount of success externally. The commercial rationale upon which the case for merger was built was proven almost immediately – right away, the new firm was winning new mandates that none of the three legacy firms would have won alone, and was able to bring to clients a compelling offering of deep and wide practice expertise across geographies. This alone – and, to some extent, the strength of the financial performance that it was securing for the firm as the year progressed, and the confidence that this conferred – was a key factor in enabling the firm to navigate with a degree of success the inevitable internal challenges that I've described. At the 100-day mark, we created and shared a short film, 100 seconds long, called *100 days, 100 opportunities*, which showcased our most exciting new pieces of work in the period since 1 May.[7]

Lawyers love being able to deliver what their clients need and to win new and exciting work. For many of them, this is their deepest purpose and the main strand of their story. And so we made sure to talk, a lot, about these successes. We pushed out 'soundbite' emails across the firm, and made opportunities for partners and others to talk about what they were achieving together. We engaged Dr Heidi Gardner of Harvard Business School, and author of *Smart Collaboration*, to work with us to help us all get better at working together. Slowly but surely, this story, its different strands told in different ways by different voices, became more and more compelling and significant. Yes, there were some inevitable internal challenges early doors – and really remarkably few – but people were overwhelmingly positive about the opportunities that the merger afforded them in terms of being able to deliver an enhanced service to clients, and they found the vision for the new firm compelling and exciting.

Tate That

All good stories need turning points and watersheds. In this story, one such inflexion point came on 16 November 2017, six months and sixteen days after the creation of the new firm, when the firm held its first all-partners meeting at the Tate Modern gallery in London.

This was the partners' opportunity to tell each other the story so far – reinforcing the rationale for having created the new firm in the first place and weaving in everything that they had learned and achieved together in their first six months – and then to draw breath and face the future together, ready to create the next chapter.

The venue could not have been better – impact, modern vibe, blank canvas – and the leadership team created an agenda which would allow for celebration, inspiration and lots of networking and discussion. The team had also created a draft strategy statement for discussion on the day, and had spent time carefully distilling and articulating the firm's purpose and vision onto one page, reflecting everything the firm had learned and created over the past six months, and everything people had spoken about, in the Open Door 2.0 sessions and more generally.

There was a deep pleasure and satisfaction in seeing how the story that was there back in the Information Memorandum was still there, but developing and maturing. There was a thrill in getting to the point where the leadership could articulate the whole thing, on one page, in big, confident letters, and with crystal clarity:

We are here to anticipate and create sustainable and rewarding futures for our clients, our people and our communities.
::

We share a vision to build a new kind of future-facing global law firm- the standout choice for clients and top talent; progressive,

agile, rooted in sector expertise, passionate about quality, and powered by technology.
::

We are dynamic and successful, empowering and inclusive, underpinned by trust and respect.

The atmosphere that day was electric, in a deep, connected, confident sense. The leadership team delivered the keynote speeches of their careers. People were engaged, showed up, contributed, shared ideas, applauded. It always feels so good to have a compelling way forward. It really is impossible to overstate the impact that this central message had for the new firm at this point.

Purpose.

Vision.

Values.

How simple it seems when written down.

How vital it would now be to press on, and deliver.

Getting It Together

A shared sense of purpose, values and vision, and the stories which supported them and brought them to life, were both the catalyst for the CMS merger taking place at all, and then the means by which the leadership team kept the whole thing on track throughout the integration period:

- they helped to create a sense of **belonging** – this is who we are, this is what it feels like to be part of the team at CMS;

- they helped the firm to **evolve** – the leadership team were clear about the end game and so could afford to be less process-driven along the way;

- they were the means by which we could bring a degree of **certainty** – and the accompanying uptick in both comfort and productivity which certainty confers – at a time when there was a huge amount of uncertainty;

- they enabled the firm to be **agile**, because everyone knew (and shared) the endgame the firm was working towards and so the leadership could devolve decision making away from central management and get decisions made quickly with minimal risk of chaos;

- they helped people to **understand** one another, personally and corporately, because their stories and experiences complemented and corroborated one another;

- they helped the leadership team to achieve **simplicity** in a truly complex context, because they could strip away extraneous rules and processes and depend instead on a shared purpose and values to hold things together; and

- they helped us to maintain **energy**, individually and as a firm, because knowing why we are here and doing something, and feeling part of a common purpose are key sources of intrinsic motivation and resilience.

Moving On

At the time of writing of this part of the book, the firm is eighteen months old. The story of its short history together is becoming richer and more multi-faceted by the day. The firm has deepened and broadened its relationships with clients, and has been the object of good industry recognition. It had a strong outcome in terms of both income and profitability at the end of its first financial year. It achieved the anticipated savings associated with the merger earlier than expected and these have more than offset the relatively few areas where the firm made some additional spend, investing in team support and development, trading some efficiency for ease and buy-in, and so on.

Even just six months ago, the firm sometimes felt acutely aware of not having a long collective history to draw on, whereas now the firm talks about itself almost exclusively as a single firm, and consciously talks about the future more than the past. Now that the firm has walked round the full circle of the year together, there are fewer 'firsts' to encounter, a greater sure-footedness and a really solid sense of trust and mutual understanding.

The senior leadership nevertheless remains mindful – thinking about how to make sure that our integration continues, goes deeper still, and sticks. As far as possible, it is good when the wider firm is able to achieve this without looking at it directly – it is much better for the lawyers to be outward-focused and laying down history with their clients, writing their story that way, and for the conditions they need for true integration to be curated around them. Three challenges, or opportunities are ongoing in this regard . . .

First, there is more to be done to help people to navigate the new firm, at scale, seamlessly. This depends on helping them to build informal networks of the kind one takes for granted when one has been somewhere for a long time. The work CMS is now doing in this area has many strands, but all depend on storytelling and have purpose at their heart.

Second is the ongoing challenge to deliver on that 'best of three' or – better – that 'fourth way' promise. It is undoubtedly the case that when things were tricky in the early days, people's innate resourcefulness kicked in and they found workarounds for all sorts of things and then shared them with one another – *oh yes, I always have to dropkick my laptop on a Tuesday*. While it is inspiring how collaborative and smart people are in terms of solving problems, there remains a need in some areas to mature short-term workarounds into long-term, sustainable solutions. Because the leadership is supremely ambitious for the firm, it does not want people to accept anything less than excellence in any aspect of what they do. The SPIT team has been a godsend in this regard – methodically working through the issues in order of priority. The substantive improvements that have resulted, as well as the demonstrable commitment that the senior partner members of this group have shown to making things better, are also being woven into the firm's story.

The third opportunity, perhaps the highest and hardest, is to continue to create a context within which true, deep relationships can grow between colleagues in the new firm. While to some extent this cannot be forced or rushed (and I for one need to resist the temptation to keep pulling up the plant to see if it is growing), my sense is that we are increasingly seeing new connections develop and deepen. It is not possible to replicate or replace in a few short months the ties that have bound colleagues together over twenty or thirty years. Nor would the new firm want to – those ties are precious. What is possible, once again, is to include and transcend. Keep those relationships and loyalties intact and build new ones. Role model inclusion. Very deliberately give people opportunities to work and socialise together and then tell stories about those experiences – the deals done, problems solved, relationships built. Publicly celebrate new relationships and tell the stories of how they came about. And keep listening, and telling new stories. This story had a beginning and a middle, but it has no end – the new firm is writing the story as it lives it.

Notes

1. Frankl, Viktor, *Man's Search for Meaning*, Simon & Schuster, 1946.
2. (and of course everyone had a chair in the end).
3. https://www.youtube.com/watch?v=BL_X7GelX5Q (barn raising scene in Witness film, 1985).
4. Eco, Umberto, interviewed in *Spiegel Online* by Susanne Beyer and Lothar Gorris, 11 November 2009.
5. 'SPIT' for short. I had wanted to call them the Process Improvement Group, but. . . .
6. Again, just experimenting with communication channels so that people could receive the information they needed in an easy and immediate way.
7. A fun and accidental joke, which likely only a handful of under 25s in the firm got, was that the music in the video was *Bills* by Lunchmoney Lewis – 'I got bills I gotta pay, so I'ma gonn' work work work every day . . . '.

Part Three

Because

This Part Three of the book is about practical application. We have explored some of the latest academic thinking around change, and some of the theory around purpose, values and the role of story-telling. I have then told you the story of the CMS merger: one story of one particular change journey, in one sector, at one time. In this part, we will draw on both of these to move towards specific, actionable ideas and approaches – in part by hearing a little from others in other sectors about their own experiences of change, and in part by interrogating more of the research and literature.

The central thesis of this section is that an approach to change which is centred on purpose and values, and relies on the creation and telling of stories can help organisations to rise to some of the biggest challenges and opportunities that periods of profound change present. In particular, it will help organisations with their **BECAUSE** – to build **b**elonging, take an **e**volutionary approach to strategy setting, build **c**onfidence, develop **a**gility and **u**nderstanding, create **s**implicity and foster **e**nergy.

The following chapters take each of these concepts in turn:

Belonging – creating a sense of belonging and buy-in; how to help people build relationships, feel at home in their new environment and develop new a multi-layered sense of identity;

Evolution – taking a dynamic and emergent approach; how to deal with uncertainty and ambiguity, how to lay the road and ride it at the same time; knowing when to be pragmatic and when to stick to core principles;

Confidence – creating confidence and a degree of certainty in an uncertain world at a time of great change; transcending and including, managing conflict, building courage and staying playful;

Agility – staying fleet of foot and getting even fleeter, being responsive and taking advantage of opportunities, while also maintaining cohesion and building trust;

Understanding – communicating enough, not too much, effectively and inspiringly; how to listen and respond and help people learn about, navigate and deliver the business;

Simplicity – helping overwhelmed people to handle change; why good stories bear repeating, and repeating, cutting through red tape and staying focused on what matters; and

Energy – maintaining energy and momentum, building and preserving resilience; figuring out what makes people tick, and the importance of wellbeing.

Chapter 9

Belonging

*For the first time in my life I saw the truth as it is
set into song by so many poets, proclaimed as the
final wisdom by so many thinkers. The truth – that Love,
Meaning and Connection are the ultimate and highest
goal to which man can aspire.*

Viktor Frankl[1]

There's a scene in the film, *The Devil Wears Prada*,[2] in which Emily (Emily Blunt) tosses a fat black binder full of photographs onto Andy's (Anne Hathaway's) desk and tells her she needs to memorise the names and faces of everyone in the binder within a few hours.

The scene then cuts to a charity gala event later that evening, in the course of which Andy leans over and whispers into her tyrannical boss Miranda Priestly's (Meryl Streep's) ear the name and backstory of a particularly prestigious guest who is approaching, thereby saving Miranda from embarrassment, outshining her co-assistant Emily, and securing a stellar future for herself.

This scene played on loop in my head in the few weeks running up to the CMS merger on 1 May 2017. I was carrying around in my handbag a dog-eared stapled-together bundle of photographs of all of the partners in the new firm, and poring over it whenever I had a few minutes to spare – on the train, in the queue for coffee, at bedtime – 'this is Phil, a partner in the corporate practice, specialising in technology . . . this is Susie, a litigator, specialising in fraud . . .'

In my case, the reasons were less to do with embarrassment, or meteoric career trajectories and all about being determined to ensure, in as many small ways as possible, that as many people as possible would feel, as soon as possible, as though they belonged in the new firm. I wanted to be able to extend people the courtesy of greeting them warmly by name as I passed them in the corridor.

What Is Belonging and Why Does It Matter?

A mountain of research which spans almost two hundred years and looks across multiple cultures and contexts demonstrates beyond doubt that the single biggest predictor of our health, happiness and longevity is our sense of connectedness to one another.[3] Read that again. More than genetics, diet, exercise, choice of career, where we live – the *single, biggest predictor* of our health, happiness and longevity as individuals is not a factor which pertains to us as individuals at all – it is about community. We need to feel connected to each other. We need to feel, in other words, as though we belong.

In a work context, our sense of belonging is rooted in myriad small things – the building we turn up at every day, our walk from the station to get there, the name above the door and on our headed notepaper, the people around us who we've known for twenty years and who know us, and our dog, and the fact that we get grumpy if Spurs lose. It's in the smell of coffee on the client floor, the annual band competition we run, the jargon and acronyms we use.

Any change which challenges our sense of belonging presents a threat. Our brains process social pain using the same systems as they use for physical pain – to the brain there is no difference.[4] Rejection and exclusion have even been shown to reduce IQ.[5] And few changes present as big a potential threat as a merger. In our case, we were changing the firm's name, and the colleagues people would be working alongside every day. For many people, they would be coming to work in a different place, at a different desk, on different IT systems, using different jargon to participate in different systems and processes. We had no shared history or rituals, no in-jokes.

This profound challenge to belonging and its implications for wellbeing matters for its own sake: we need to enable organisations to respond to their volatile and uncertain market environment and make changes without exposing all of their people to an undue threat to their health, happiness and longevity. But it also matters, in commercial terms, to the health of an organisation's business, and to an organisation's ability to embed and secure the changes it is trying to make.

What Happens to Our Sense of Belonging When Things Change?

So, what is going on for us as individuals when circumstances change such that the things that make us feel we belong fall away? The answer lies deep in our limbic brains. When we are presented with some new circumstances, our newer, outer brain – the neocortex – does a really good job of grappling with all of the factual information pertaining to those new circumstances. The neocortex can digest all of the facts and understand why the change proposed makes good sense, intellectually. But the limbic brain – our older, faster, animal brain – will overrule all of those rational conclusions based on analysis of the facts in favour of what 'feels right'. The limbic brain deals in loyalty and trust, and it's the one in charge of decision making. Thinking again about David Rock's SCARF model, introduced in Part One,[6] we can see that when we are not sure we belong somewhere, we can experience a challenge to our sense of status, our sense of certainty and our sense of relatedness to other people. When we talk about the need to win 'hearts and minds', we are really talking about the need to win over the limbic brain as well as the neocortex.

At CMS, prior to the merger, this was one of the most commonly expressed concerns across all three legacy firms – people really valued the intimacy of their own firm and worried that a much larger firm would be a cold, impersonal place where nobody would know them or their history; a place where their old stories would drift away, where they would feel lonely. The precise dynamics of this issue varied according to practice group or sector – in some groups there would be almost equal representation from the three legacy firms in the new firm; in others, one or two individuals were going to be joining an already well-established team. The team at CMS took various steps, as outlined below, to help people to build relationships in advance of the merger. They also used some more practical measures, such as ensuring that seating plans were arranged such that everyone was within sight of someone they knew.

Getting the limbic brain on board depends in no small part on the extent to which it feels as though it belongs in the new circumstances, and so a sense of belonging really matters if you want to get people on board with change.

In *Start With Why*, Sinek neatly illustrates how our sense of belonging can sometimes be incidental, and sometimes central, depending on circumstances, and how the facets of our identity that most matter to us can shift. Sinek describes meeting a fellow American on a bus in Australia, and the feeling of connection and belonging that the chance meeting engendered so far away from home.

The natural 'default' for people – what makes their limbic brains immediately feel better – is to gravitate towards old social networks and old ways of working, In order to ensure the best of success when bringing groups of people together, organisations should address this by doing two things: first, respecting those old bonds and patterns and identities and enabling them to continue for a period, while people need them; and secondly by, pretty quickly, engendering such a sense of belonging to the new organisation that new networks and identities and ways of working can develop which transcend the old.

How to Build Belonging

There will be, I believe, for every organisation some fundamental aspects that must be in place before the deeper work of building belonging can begin in earnest. The precise list will vary from organisation to organisation. For CMS, for example, the list included co-location in London, and operating under a single brand on a single system. The absence of either one of these would have led to such a gulf that the threads of new connection could not stretch across it.

The good news is that, *provided these fundamentals are in place*, making new connections in an organisation is not so terribly difficult to do, and the act of doing so actually strengthens us and builds resilience. Interpersonal trust begins to build rapidly after only a couple of personal interactions and, at a neurological level, the anticipation of a rewarding relationship with someone new, or the anticipation of learning about a new way of doing things, gives us a dopamine hit, which may in turn make us more prone to take a risk and reach out to someone new, or try something different, next time too – a virtuous circle.

In my experience, there are broadly three areas which it can be productive to focus on in terms of engendering a sense of belonging, perhaps particularly in the context of merger or when otherwise bringing groups of people together:

- finding different ways to help people to connect with each other and build relationships;

- helping people to feel at home in their physical environment and in the routines of their working day; and

- accelerating the creation of a new sense of identity.

In all three of these areas, purpose and stories play a central role, as we shall see.

Connecting and Building Relationships

Depending on the particular context, scale, history and so on, it will be appropriate to take a more or less structured approach to encouraging people to meet one another and begin to build relationships. At CMS, there were already lots of connections, professional and social, between people across all three firms, and so lots of ad hoc discussions and meetings were already taking place organically, and there was huge value in the unexpected connections that people made in this way. But it was also important in CMS's context to make sure that everyone was included and nobody felt overlooked or sidelined, and so the team also had a big master chart that looked like an enormous family tree where they could check off every permutation – *Have the practice group leaders connected with every partner in the future team? Have sector leaders met with each other? Has every associate had the chance to meet someone from another firm?* etc. This sort of structured approach does not, of course, speak to the quality of any of the interactions, but at least it establishes a bedrock of relationship which can be built upon later.

Connecting at the Centre

CMS devoted considerable time pre- and post-merger to building the relationships between the future members of the new firm's Board. This was a smart move – it was important that there was a sense of trust, intimacy and belonging right at heart of the firm's structure from the outset – both because this would enable good decisions to be made, and quickly; and because the relationships between this group of people would be the role models and catalysts for the relationships being established across the rest of the firm.

In this context, it is important for organisations also to consider what, from a governance and systemic perspective, they want their boards and senior teams to do and be, as this will drive, and by driven by, how these central senior relationships develop. Subject to ensuring good governance, there may sometimes be a case for flying in the face of received wisdom in order to achieve a particular end. So for example at CMS, the senior management teams of the three legacy firms were absolutely clear that a larger-than-usual board would be the best structure for the new firm given its own particular story. Similar thinking can prompt organisations to appoint co-leaders to lead parts of a business following merger. It was of paramount importance to the CMS leadership that the board should help to build buy-in and underpin the collaborative, transparent and democratic culture that the new firm wanted to establish, even if that meant sacrificing a degree of efficiency around decision making by virtue of the sheer number of voices around the table. In fact, eighteen month on, the 'too large' board at CMS has proven to be a highly efficient, open and trusted body where robust discussion is embraced.

In my experience, it can be helpful as part of establishing a new board or senior team to bring everyone together explicitly to talk through their roles and responsibilities around leadership, stewardship and risk management and to agree how the team wants to work together, and the sort of culture they should emulate. This ultimately helps with governance and decision making, but it also helps to engender a sense of belonging amongst the most senior people in an organisation.

Connecting on the Ground

As outlined in Part Two, on Day 1 of the CMS merger, everyone in London had a name badge on their desk and was strongly encouraged to wear it, and to be proactive about introducing themselves to one another. The coffee vouchers, coffee lotteries and huddles gave people some loose structures within which they could begin to interact.

In the summer of 2017, once the immediate dust had settled, and just as people were losing their name badges and getting a little embarrassed about introducing themselves yet again to someone they may have already met last week, the team at CMS stepped it up a bit. You read in the merger story in Part Two about quizzes, speed-dating, breakfasts, seminars, flashmob choirs, sports, volunteering, sounding boards and think tanks . . . all of it designed to appeal to as broad a group as possible, to draw people in, and to begin to create a sense of purpose and lay down stories. Every organisation will want to consider what sort of interactions will work best for them – there is no single right answer, and no shortage of possibilities. At a team level, various practice groups and sectors and networks at CMS organised offsite events – team-building weekends in Brighton, walks in the Peak District, workshops on resilience – all of it targeted to particular audiences and with a different emphasis depending on the context, but all contributing, too, to building this sense of belonging. One newly created team at CMS created its own team logo, and had it branded onto brightly-coloured leather satchels and document wallets for all team members.[7] I bought my own team matching friendship bracelets with brightly-coloured threads, engraved with 1 May 2017, and 'Only the beginning' – partly as a thank you, but also to mark them out as a team. We all wore them with pride.

Connecting Remotely

Building a sense of belonging in CMS London presented challenges largely around scale, complexity, intimacy and logistics. Building a sense of belonging for the parts of the firm outside London, from Beijing to Rio, presented a different set of challenges – around relevance, inclusion, and feeling valued and understood.

It sounds blindingly obvious, but the two most important lessons I have learned about building a sense of belonging across geographies and markets are, first, that it takes time – usually longer than it takes 'on the ground' by virtue of sheer logistics and because of cultural differences – and, secondly, that it is important that everyone is oriented to 'moving towards' rather than 'moving away' – actively going out of their way to look for opportunities for each other and to work together. It can feel forced, time-consuming and awkward, but it absolutely always pays off. Perhaps there's a third lesson hidden in there about the value of meetings. All of this involves lots and lots of face time – inefficient, repetitive and expensive – but there really is no substitute for having people meet one another face to face and find their common ground.

Tribe-making

Slowly but surely, organisations and leadership teams create a sense of belonging by reminding people – sometimes through explicit words and 'telling', sometimes through the lived experience of people working together to create something – of the purpose of what they are doing together. The role of leaders is both to tell the stories, and to draw people into those stories so that they become a part of what is going on – bonding with the storytellers, and with those new people around them who are also joining the story. It is about laying down shared experiences and beginning to bear witness to each other – 'I was there when . . .' 'Do you remember when. . .?'

All of us want to feel as though we're not alone. We also want to feel special – to understand our unique part in the story and to know we matter. In his book *Tribes*,[8] Seth Godin says that a group needs only two things to be a tribe – a shared interest, and a way to communicate. People, he says, want connection, growth and something new. Godin references Senator Bill Bradley, who says that a movement has three elements:

- a narrative – a story about who we are and the future we're trying to build;

- a connection between the leader and the tribe, and among the members of the tribe; and

- something to do – the fewer limits the better.

At CMS, these three ideas – a story, connection and something to do – were all bound up in the purpose of the new firm, and in the stories they were creating and telling throughout that first summer. In a nutshell, they were creating a new tribe. I saw the firm make very good and very deliberate progress on the first two of these throughout the earliest weeks following the merger, but when groups and teams found that third element, their 'something to do' together, the sense of belonging increased exponentially. In the context of a law firm, the 'something to do' is almost always work for clients – in my experience, there is no better way to build a sense of belonging to a firm, and esprit de corps between new colleagues, than by finding opportunities for them to work together, bringing their collective expertise to bear to do the thing that lawyers love best – namely, delivering an excellent result for clients. For other organisations the 'something to do' will be different, but the outcome will be similar – a well-bonded team with a strong sense of belonging to the organisation.

A House Is Not A Home

The second area that the team at CMS focused on, which relates to my second point above around focusing on the physical environment, was dailyness – what does it actually feel like to turn up at work each day, and do your job? The team looked at the systems and processes that people would have to grapple with every day and at the physical environment. As human animals we feel safer when our surroundings are familiar; happier when they are light and pleasant; better able, as a very practical matter, to build relationships and community when we can actually find our tribe.

Cannon Place is in the heart of the City of London. Originally built with a bank tenant in mind, the floorplates are huge; designed to accommodate a trading floor. On any given day, it's easy to clock up a healthy 10,000 steps without leaving the building. The legacy CMS firm moved in to the building in the summer of 2015, and took advantage of their move from a tired old building in the shadow of the Barbican to invest in the best tech, and to create an office full of light and bright colours and spaces for people to collaborate. The CMS office won accolades for being the best of its kind in the City, and people were quickly happy and settled in their new space. In advance of the merger the new firm took two additional floors with views of St Paul's, the river, the London Eye, the Tate Modern – the whole sweep of the City for a new City powerhouse.

If the advantages of a big and beautiful open-plan office are obvious – collaboration, buzz, a sense of scale and industry, efficiency – then so, too, are some of the challenges: it can be difficult to find your way around or to find the people you're looking for, and it can feel impersonal and daunting. For some in the newly merged CMS, the whole concept of open plan was new. With this group, the team took extra care, making the case and sharing experiences and the latest thinking around the neuroscience of attention; the social science of interacting in and across groups and teams; and the very British art of knowing when to interrupt someone and when not to. Everyone in the new firm was invited to have a tour of the building in advance of 1 May, and on Day 1 there were 'you are here' maps and the team of lime-clad helpers all over the floors, and guides to the building and pop-up 'open-plan etiquette' cubes in all the goody bags.

If more proof were needed that the physical environment matters deeply to a person's sense of belonging, it was there in bucketloads by the end of that first day at CMS: new desks across the building, many of which hadn't even been built three days previously, were bedecked with family photographs, children's drawings and favourite mugs.

It felt important at CMS that absolutely everything should bear the brand and identity of the new firm – from the email boilerplate, to the labels on sandwich platters in meeting rooms, to pens. The firm ran a photography competition ahead of time and collated all the entries into huge map-of-the-world wall montages which were made into a feature wall in almost every office – adding colour, fun and a really important sense of engagement and belonging for everyone who pored over it in a coffee break to find their dog on a beach a little west of Fiji, or their wedding day adorning the tip of Manhattan Island. People bought their coffee from Crumbs – the new coffee bar on the seventh floor that a junior lawyer had named in a competition. Again, all of these examples are specific to CMS, and other organisations will have different ideas, but the overarching point is that the physical environment can be a powerful factor in helping to establish a sense of belonging, and that little details can make a huge difference. In time, teams added more artefacts to their physical environment, making it feel more like home – house plants, cartoon sketches from the inaugural partners' conference, a neon pink 'Media City' sign on the wall beside the media teams' desks. All of these things were tangible signs of the new stories people creating together – bearing witness – 'we were there when . . .'

In addition to considering the physical environment, the team at CMS also looked at the processes and systems that people used every day. As discussed in Part Two, these were a source of challenge in the weeks following the merger, and there is more about how the team at CMS went about tackling those in the chapters to come on Evolution and Simplicity, but in the context of belonging, the priorities were twofold – first, to familiarise people with the systems as they were and to help them use them; and secondly, to ensure that people had a voice and could influence future systems and processes to better meet their needs.

The first is important in terms of belonging because an inability to engage with a particular system or process can quickly lead to feelings of isolation, and to a perceived threat, in David Rock's terms, to status, certainty, autonomy, relatedness *and* fairness. The full SCARF! Offering help – at CMS, mostly in the form of an army of lime-green floorwalkers, drop-in sessions with cake, and some online how-to's – both lowers the sense of threat by building competence and confidence, and conveys to the person in question that they matter and that their organisation cares about them.

The second aspect – ensuring that people have a voice and influence to shape systems and processes – is important partly for the same reasons, and also because it builds buy-in and a sense of collective identity. Think about your own experience. Speaking personally, as someone not naturally inclined to follow rules or processes, I am much more likely to adopt, buy into and be-patient-with-the-inevitable-imperfections-in a process or policy which I have been involved in shaping, even if only a little. At CMS, SPIT – the Systems and Process Improvement Team – was established to give everyone this chance and, again, to signal the firm's commitment to building a firm together that would work for everyone.

Identity

The third area, then, to focus on in terms of engendering a sense of belonging is to encourage and enable the development of a keen sense of identity for an organisation and its component parts. This can be complex in all but the smallest of organisations because it tends to involve layers of inter-related aspects – for CMS, individuals, huddles, teams, practice groups, sectors, offices and the whole firm. It was very important to the leadership team at CMS that people should feel able to be entirely themselves and that the firm should be an inspiring and diverse place to be. CMS wanted teams to have a lot of autonomy and the space to develop their own distinct culture, vibe and market position. But the firm also needed to develop an overarching firmwide sense of identity – to be 'CMS enough' that the whole enterprise would hang together and make sense, both internally and in the market.

Remind Me Who I Am Again?

How to go about establishing a sense of identity at an individual level within an organisation, is a slightly different point to the issue explored earlier around building relationships and ensuring people feeling known by one another. This is more about an individual's sense of self and standing, and that being somehow vested in, or at least related to, the identity of the organisation.

At CMS, the challenge was twofold. First, to give everyone the confidence that all of their expertise, goodwill and reputation would be recognised and celebrated – that, in David Rock's terms, their status was intact and not under threat. Secondly, to do everything possible to help people to evolve a new identity which transcended and included their old one, and incorporated belonging to the tribe of CMS as part of it. Stories and purpose helped on both fronts.

As to the first, creating opportunities for people to showcase their expertise – 'Meet The Partners' emails, lunch and learn sessions, guest slots in townhall meetings – enabled people to assert their identity in the new context and, even for those who did not participate themselves, conferred a confidence that this was a firm where people were seen and heard and valued for what they could bring. As to the second, speaking in terms that people could relate to about the purpose of the new firm, and sharing stories about new opportunities that had arisen for individuals by virtue of the merger made a new 'future self' tangible for people. Every-one received new business cards on Day 1, and was asked to update their LinkedIn profiles – all little tangible manifestations and encouragers of a new identity. Talking constantly about inclusion and supporting this with clear signals such as the decision not to have any strict dress code for the new firm also helped.[9]

There's No 'I' In Team

In 1992, Robin Dunbar, the anthropologist and evolutionary biologist, published the findings of a piece of research which correctly predicted social group size across dozens of species of primates, based on the size of the neocortex. Extending this principle to human primates, he predicted that the largest effective group size amongst humans is around 150 people – this became known as Dunbar's Number.

This means that if people are to establish a clear sense of group identity within a large organisation, they are much more likely to – and well advised to – identify first and foremost with a sub-group within the organisation, rather than with the organisation as a whole. This is to be actively encouraged. For many their primary 'tribal home' will be the team that they physically sit beside and work with day to day. The visual and environmental factors which are discussed above – pot plants, neon lights, cartoons – help with this, and so too do the myriad 'ways we do things around here' – weekly newsletters, banging a gong to celebrate success, going for burritos together on a Thursday lunchtime. Some will also find other groups to develop a sense of affiliation with sports clubs, choirs . . .

New-Firm-Enough

. . . This brings us to the third aspect of identity building – namely, building an identity for an organisation as a whole. One of the things that made CMS's inaugural partners' conference at the Tate Modern so special was the palpable sense of the presence of the new firm as a firm, rather than as a collection of individuals, – 'We are here. This is who we are and what we are here for.' Everyone in the room could feel it.

At the organisational level, identity is closely linked to brand, and while this is best thought of as an inside-out relationship where identity drives brand, it is also true that positive press coverage, strong performance in league tables, the odd huge billboard in railway stations, and the challenge of spotting the three CMS-wrapped taxis pootling around London, all contributed to building a sense of identity in the early days of the CMS merger.

There is also a leadership and governance angle. The most senior leaders and the board have a key part to play shaping, articulating and role-modelling a common sense of identity for an organisation. Here, again, the value of a large buy-in board and/or extensive co-leadership following a merger is apparent: you need a large number of senior people with strong and influential affiliations now deeply and visibly a part of the organisation's future, and bringing their perspectives and stories to bear to shape its identity. The language this group uses – *we* not *them, future* not *past, hope* not *loss* – the stories they tell about what they witness happening in the here and now, the priorities they set . . . all of these have a profound influence on how an organisation shows up and how it thinks and talks about itself.

Underpinning all of this, though – the events, the advertising, the inspirational leadership and the stories – must lie a common purpose, and shared values. Following their merger, the team did a lot of thinking and a lot of talking in order to figure out what it meant to be 'New Firm enough'; what it would take to pull together the rich heritages and cultures of the three legacy firms and to bond the new firm strongly together while also allowing all that space and freedom for layer upon layer of identity to grow. The answer which emerged loud and clear was, first, that everyone had to be pulling together towards a common purpose – albeit with all of the myriad nuances and variations and elaborations we have explored previously. That purpose was as set out in Part Two, namely to anticipate and create sustainable and rewarding futures for their clients, their people and their communities. The second part of the answer was that everyone in the firm had to be doing this in a way which embodied the new firm's deepest common and distinctive values which, again, emerged from months of storytelling and listening, namely: we grow, we are resilient, we engage, we include, we trust and respect, we collaborate, we innovate. The purpose and values which the new firm expressed were not, in that sense, 'new' at all – rather, they were very much rooted in and reflective of the strong orientation towards a holistic and inclusive approach to business that all three legacy firms were already deeply committed to.

This is consistent with Tristram Carfrae's observations of how things work at Arup – an organisation with an outstanding reputation for both a strong and clear common purpose and high levels of individual and team autonomy:

> Working down through all the layers of identity, once you get to the bedrock of our aims, and to belong to Arup, you basically have to agree with them. We aspire to be a truly inclusive organisation, but in this respect at least we are actively discriminatory – we demand alignment with our aims, and the intrinsic motivation to pursue them.

This is what an organisation's identity rests on – what it is here to do, and how it will do it. CMS incorporated these concepts – its purpose and values – into its approach to performance and reward, its communications strategy and its development programmes. They became the non-negotiable bedrock across all parts of the business, whatever layers of local practice group, sector, geographic or sports-based identity were built on top. The leadership at CMS wanted to empower individuals and teams to be what they wanted to be – and what their clients and markets need them to be. They did not need, or want, everyone to participate in the same way, but they did need everyone to align with the firm's purpose and to engage, include and collaborate in line with its shared values. In essence, the firm needed everyone to adopt the 'we grow' mindset and to be thinking less about what they could get, and more about what they could create together.[10]

Tristram Carfrae again:

We're trying to create a carpet . . . a tapestry . . . a patchwork quilt is better. Yes, a patchwork quilt that everyone can be part of. We actively want the patches to be as different as possible, but they have to be stitched together. It's our aims, our purpose, that stitches us together.

Conclusion

To conclude, then – engendering a deep sense of belonging is absolutely vital for individual wellbeing and corporate performance, particularly during periods of profound change which pose a threat to our sense of self and safety. Three aspects are in my experience particularly productive to focus on – building connectedness and relationships; creating an environment in terms of physical surroundings and daily routine in which people feel safe and authentic and can thrive; and building identity, for individuals, teams and the organisation as a whole.

There are lots of practical things organisations can do to tackle all three aspects, and the details will vary according to context – but at core, for all organisations, a clear sense of common purpose and some shared values are, I believe, vital. Stories are one of the most potent tools to help with building connectedness and identity in particular, because they engage people at the deeper emotional level at which issues of belonging and identity operate, and build a richer and more vibrant context for people to aspire to be part of.

As we shall see in the next chapter, stories, purpose and the sense of belonging they engender will serve organisations well when they encounter a set of circumstances which throw the best-laid plans and any traditional 'strategy' up in the air, and demand that they think on their feet and take a more evolutionary approach.

Notes

1. Frankl, Viktor, *Man's Search for Meaning*, Simon & Schuster, 1948.
2. *The Devil Wears Prada* movie, 2006, based on the 2003 novel by Lauren Weisberger.
3. See for example, Diener, E., and Chan, M.Y., 'Happy People Live Longer: Subjective Well-Being Contributes to Health And Longevity', *Applied Psychology: Health and Well-Being*, 3(1), March 2011, and Capaldi, C., Dopko, R., and Zelenski, J., The Relationship Between Nature Connectedness And Happiness: A Meta-Analysis, *Frontiers in Psychology*, 5: 976, 2014.
4. Scarlett, Hilary, *Neuroscience for Organizational Change: An Evidence-Based Practical Guide to Managing Change*, Kogan Page, 2016.
5. Baumeister, R.F., Twenge, J.M. and Nuss, C., 'Effects Of Social Exclusion On Cognitive Processes: Anticipated Aloneless Reduces Intelligent Thought', *Journal Of Personality And Social Psychology*, 2002, 83(4): 817–827.
6. Rock, David. *Your Brain at Work*, HarperBusiness, HarperCollins, 2009.
7. www.zatchels.com - they are gorgeous!
8. Godin, Seth. *Tribes – We Need you to Lead Us*, Piatkus, 2008.
9. The dress code simply says that people should dress to 'inspire confidence'.
10. Dweck, Carol S., *Mindset – Changing The Way You Think To Fulfil Your Potential*, Updated Edition, Robinson, 2017.

Chapter 10

Evolution

You start a painting and it becomes something altogether different. It's strange how little the artist's will matters.

Picasso[1]

You Say You Want an Evolution . . .

Strategy is dead! So says pretty much every business academic worth their salt these days. These days, it is all about VUCA – that is to say, our world is so volatile, uncertain, complex and ambiguous that it's a wonder we can put one foot in front of the other. I'm overstating of course, but it is certainly a while since any of the organisations I am closest to has undertaken meticulous months-long research and analysis and drafted a detailed five-year plan, running to hundreds of pages – an approach that was standard practice only ten years ago. This is primarily because we know that the pace and nature of change in our market and environment is such that we have to be able to change direction on a hairpin. The financial crisis in 2008 put paid to any complacency on that front.

That does not of course mean that every organisation just wakes up and reinvents itself every morning. However chaotic circumstances may feel on occasion, we are none of us, really, just making things up as we go along. Clearly, then, there is some sort of balance to be struck between a three-hundred-page gospel and some scribbles on the back of an envelope. But how to strike that balance? How much planning and articulating and fettering and tethering is 'enough'? How can an organisation remain open to possibilities and awake and reactive, and yet crack on with intent, focus, and without mayhem? And why? Might there even be an upside to this sort of evolutionary approach?

Certainly, this way of working is not as scary as it sounds. Chances are, lots of us are doing it anyway. I challenge you to scrabble around in an old filing cabinet and pull out the last five-year plan your organisation produced. Flick through the pages and make a quick assessment as to how much of what was planned has come to pass, or has been delivered. Chances are it will be a bit of a mixed picture. Chances are, too, that it doesn't matter much; that the confidence and momentum you have experienced in taking your business forward day to day does not necessarily bear much relation to any sense of comfort and confidence you may have felt when that glossy brochure landed in your hands, hot off the press. If, then, the comfort and sense of control that a detailed plan brings is to some extent a false sort of comfort, then we can afford to loosen our grip on them a little. Not to let them go completely, perhaps, but just to loosen off.

This chapter explores the case for an evolutionary approach to strategy setting. Beyond necessity, why does it matter? What might it better enable organisations to do or be? It then explores what needs to be true, and what conditions need to be in place, in order for evolution to happen in an organisation, and shares some reflections on how to build a culture that supports an evolutionary approach – hopefully without scaring everyone half to death.

Turning Points and Watersheds

The leadership team at CMS had a clear strategy and detailed plans for the delivery of the merger. Of course it did! Lots of them. I have mentioned the Information Memorandum which set out a compelling future story and vision for the firm, and plenty of detail around specific opportunities and priorities. I have also talked about the planning and strategising the leadership team did in relation to the implementation of the merger itself – the various workstreams, all lined up like matryoshka dolls. I may have mentioned the mindmaps, and the Gantt charts. So neat. So colourful. So reassuring.

And yet, of course, events never unfold exactly as planned and the team at CMS came to understand the value of being more evolutionary in terms of both the overall strategy for the new firm, and the plans for integration. This was the inevitable outworking of the firm's open, empowering and inclusive approach – people were engaged and bringing new ideas and new perspectives to bear.

One of the things that always strikes me in relation to change projects – perhaps in relation to all of life, if we are paying attention – is how a few factors or events will come together, apparently coincidentally, that begin to signal a particular theme, or suggest a change of direction.

One illustrative example of this came around nine weeks in to the CMS merger, just as the firm completed the second month's financial close. Some people were still finding unfamiliar systems time-consuming, but many more people had engaged with the process, in part due to our more hands-on approach. In the same week some noise arose around secretarial resource levels in the evenings and a routine IT committee meeting uncovered an issue. On three separate occasions that week, a member of the management team approached someone in the practice for an ostensibly innocuous discussion and found themselves faced with an outburst of frustration. The summer was fast approaching, and the leadership team really wanted to get to a relatively settled and happy sense of business as usual before people went on holiday.

Clearly, the senior leadership team needed to respond, and rather than do so in a reactive way, the team concluded that they needed to change their outlook. Like Maverick in *Top Gun*, they were 'holding on too tight' and needed to be more expansive and creative – even to be willing to run a marginally less efficient or fractionally more expensive operation, for an interim period, in pursuit of the greater good of alleviating difficulties and enabling the compelling rationale for the merger to take root in the lived experience, hearts and minds of people across the business.

The leadership team pulled together a group of people from across the business – finance, risk, secretarial, HR – people who by this point looked like a bunch of battle weary pirates – and asked them 'what if' questions. The team pieced together a list of immediate actions they could take. Some required software tweaks, some additional resource. They made an assessment of how much inconvenience they were storing up for themselves for the future date when they would want to re-visit the need for efficiency, consistency and risk management, and shortlisted those actions which made the cut. They finalised the immediate actions list and sense-checked it with the operational business leaders. The leadership team then got a note out to everyone explaining what was proposed, and cracked on with implementing the promised changes.

This all took time that people had planned to spend on other things. It meant taking a bit of a risk, and swallowing some of the pride of ownership that some people had in the efficient processes and systems they had worked hard to design and implement. But it was important and, on balance, well worth it – not just because it made some key processes objectively easier for people at a difficult time, but also because it signalled the willingness of the management team to listen and respond – and so also gave key stakeholders a renewed sense of ownership and influence and kept them more engaged in the wider integration process.

What Does an Evolutionary Approach Bring to Business?

The above example from CMS experience illustrates four different advantages which I believe a more evolutionary approach to strategy setting and prioritising can bring to an organisation. These are as follows:

- the ability to move swiftly – to be decisive and not allow a vacuum to develop, and so to bring certainty and confidence;

- the freedom to move in a manner which bests suits a particular complex context at any given particular time, while still being confident that the desired outcome will be achieved;

- the empowerment and increased engagement of everyone in the organisation; and

- better-quality decisions being made at the coalface by those best qualified to make them.

Let's now consider each of these in turn.

Go Fast, Always

This first one is a little counter-intuitive – surely, if one is being evolutionary in one's approach, one is moving slowly – taking time to seek input, reflect? Not necessarily. The CMS example demonstrates how, while inhabiting an evolutionary mindset, the leadership team could seek input, reflect and respond in a rapid and continuous way to deliver huge benefits.

Careful speed is vital for almost every organisation, in the execution and delivery of their day-to-day businesses – the need to take new products to market first, to create efficiencies, to respond to client demands, to test ideas and move on, to 'fail fast'. By operating in a way which does not depend, first, on the development of a detailed plan and then, secondly, on the assessment of any proposed action for fit against that plan, organisations develop a bias towards action, and adopt a way of working that is actually all about doing things. In time, processes and systems will accrete around this approach, and that is well and good, provided they don't slow the process down. At times of great change, doubt and uncertainty creep in to any hiatus. Moving fast closes those gaps, and helps to create confidence and momentum.

One Destination, Many Paths

One of the most short-term-disconcerting and long-term-helpful things a coach ever said to me, when I was in one of my more frenetic over-planning phases was:

The path you can see isn't the path.

I've googled this since and she either made it up, in which case she should start a career in making fridge magnets, or it's a paraphrase of this longer quote by Joseph Campbell, an American professor of mythology:

If you can see your path laid out in front of you step by step, you know it's not your path. Your own path you make with every step you take. That's why it's your path.

This is a theme that we will return to several times in the course of this book, but it first shows up here in the context of the advantages to a more evolutionary approach to strategy setting and execution because it makes clear an important distinction between a path and a destination. Being evolutionary in approach does not mean being uncertain as to where you are going – though the details may develop and shift over time – but it does mean that you can be very flexible as to the path. This is good news during periods of change in particular because it means you can 'pick your battles', and constantly assess what is most important at any given time, ensuring you are building consensus and taking opinion with you.

Process is a false friend in times of uncertainty, I have found, because its working depends on context, and as soon as circumstances shift – which, as we've discussed, is almost constantly during periods of change – aspects of the process will founder. Story, by contrast, is an approach which is focused on underlying purpose and ultimate outcomes, and can help to keep people and ideas tethered yet flexible.

In the example above, the leadership did not abandon any of its planned outcomes – but they adjusted the length of the path to get there, and the milestones along the way. They listened and responded to the higher priority at that particular time for their particular firm, around maintaining morale, engagement and a singular focus on clients, while still keeping the endgame in sight.

Teach A Man to Fish

The third advantage of an evolutionary approach is that it empowers people in the organisation and increases engagement – because there is a clear listen-reflect-act loop in place, and people can see the difference and impact their ideas and input can make.

If every detail has been worked out in advance, the role of everyone in the business simply becomes one of execution, and while there is motivation to be had in doing a really good job of delivering against a plan, there is much more to be had from feeling that one has had the opportunity to influence the plan, even in a very small way. There is a good body of research evidence that supports this, including, for example, a brilliant series of experiments recounted by Professor Dan Ariely in one of his TED talks and accompanying books.[2]

In one of these experiments, participants are asked to make origami birds, and then to put a value on them for sale. The first important finding, which speaks to motivation and meaning in our work generally, is that the people who have the satisfaction of having made the origami birds themselves, following the instructions, place a higher value on them than do the bystanders who have not been involved in making them. The second finding, though, even more relevant for our present purposes, is that the people who place the highest value of all upon the birds are those who have made them *without* following any instructions at all – even though, to any objective eye, these birds are rubbish.

People like to have a say in how things are done. They like to feel that their ideas and suggestions have an impact. This is perhaps particularly true for lawyers, who tend to have a very high need for autonomy, and need to feel that that autonomy is recognised and respected by their leaders. During the CMS integration, I saw that that the more independence and empowerment the leadership team could grant to people, the better the levels of engagement with the issue at hand.

Empowerment is an overused buzzword in the corporate world, and it has become a little less potent as a result. It can feel odd and frustrating sometimes to be required to actively empower someone. Surely we are all born empowered, and with a natural bias to action? And no one has enslaved us or denuded us of power? True, but people quickly become used to the sort of command-and-control structure which prevails in so many industrial-age hierarchical organisations. They become passive and unquestioning. Organisations and leaders that want to empower and engage their people, and thus be able to tap into their expertise, need to be very explicit that this is what they are doing, and one of the best ways to be explicit is to ask people for their views and to let those views very obviously influence decisions and outcomes. Organisations also need to think hard about their structures and processes (see Chapter 14 – Simplicity) because simply exhorting people to be proactive and take responsibility and so on will make no difference at all if the whole system reinforces command-and-control.

Coalface Mastery

This brings me to the final advantage of a more evolutionary approach to strategy setting, which is that organisations that adopt this approach will find themselves the beneficiaries of objectively better decision making – dodgy origami notwithstanding – because those who are closest to the detail, the client, the product, the market have the autonomy to make operational decisions which are aligned with the organisation's purpose, without being constrained by an overly detailed strategic plan.

Christoph Stettler, beautifully spoken and razor-sharp, is Chief Marketing Officer at Swiss software company ELCA. Prior to that, he spent five years at Franke, the kitchen manufacturer, latterly in a strategic marketing role. Across these very different industries, and drawing on his earlier experience in private equity, management consultancy and law, Christoph sees many common threads both in terms of the thinking errors that organisations tend to make when it comes to strategy setting, and in terms of what 'good' strategy done well looks like. Christoph says:

> If you can get lots of human beings all motivated together to pursue a goal which is roughly right – without worrying about the details – that is worth a lot . You end up in a better place than you otherwise would have because motivated individuals with space and discretion can fill or correct strategic gaps or errors at the operational level.

By way of example, Christoph recalls a period during his time with Franke when the organisation was making a shift from being primarily a supplier of individual components to being a supplier of integrated kitchen solutions. In order to make that shift, it would be necessary for the whole company to be able to produce and supply electrical products – because a kitchen needs a hob and a hood, as well as a sink. The Southern European part of the business had been producing electrical products for some time, but this had not been part of the core business of the Northern European division, and there was some resistance here. In reaching a solution, the constructively critical spirit and detailed input of representatives from the Northern division, together with the expertise, passion and receptiveness of the representatives from the Southern division, enabled significant improvements to be made to the product offering. Christoph attributes this outcome to the very clear shared purpose across every part of Franke to 'Make It Wonderful'. Much more than a slogan or brand positioning, this is a unifying purpose that works at the individual, team and corporate level across Franke to incite, engage and empower people to come together to produce great solutions.

How to Build an Evolutionary Organisation

So, if the prize is a fast-moving, flexible approach, empowered people and better-quality decision making – what does a winning organisation look like? Organisations that are truly evolutionary in their outlook are strikingly focused on outcomes rather than means. They focus on the big picture. They favour values, deeply rooted and widely lived, over rules and policies. They are experimenters – they try new things and don't worry too much if some ideas don't work. There are, I think, three guiding principles or approaches which underpin these characteristics, namely:

- vulnerability

- hope

- a practice of stop, look, listen and think.

Vulnerability

Research indicates that being able to take a more evolutionary approach, and to let go of the certainty that a firm plan confers, requires a degree of vulnerability. This is because it requires us to admit that we don't know the answer to everything.

Brené Brown has researched and written extensively on vulnerability and its potentially transformative power for individuals, including in a work environment. In her 2012 book *Daring Greatly*,[3] she references a compelling body of research on the role of vulnerability in leadership. Leaders who show vulnerability are perceived as courageous and can inspire others to follow suit. in addition, the sheer act of admitting to not having all the answers frees others up – gives them the space, opportunity and agency to create, take initiative, innovate and experiment.

While we might understand this intellectually, it feels deeply uncomfortable, both personally and corporately. We still inhabit a paradigm which equates leadership with expertise, problem solving, having all the best ideas, and being directive. What's more, if we are to retain ultimate responsibility as a leader – and I believe we should – then it feels much more comfortable to bear the weight of that responsibility if one also believes that one retains ultimate control. There is also a point around a leader's responsibility as the storyteller: at times of great uncertainty, we need to be able to look to our organisation and its leaders and to believe that they, at least, know where we are all going.

But we must not equate vulnerability with weakness, vagueness or woolly thinking. Unbounded, yes – a leader simply saying, 'I don't know' could be weak. It could be indicative of a lack of direction, or even competence, and it could lead to uncertainty and a crisis of trust in the wider organisation. That is absolutely not what I am advocating for here. But it is also abundantly clear from the business and political world around us, surely, that a point-blank refusal to engage with some vulnerability presents even greater threats to trust and good decision making – namely blaming, avoiding the question, making excuses, or hubris.

By contrast, properly framed, the vulnerability of not-knowing can be a defining trait of the most transformational leadership. By 'properly framed', I mean:

- clearly placed within the context of the purpose of the organisation;

- consistent with the organisation's values; and

- positioned as part of the particular narrative within the organisation at this particular time.

A 'good' vulnerable leader is not saying, 'I don't know *why . . .*', or 'I don't know *what's important*' or 'I don't know *where we are headed*' . . . Rather, a good leader is asking 'how' questions:

This is a new and complex challenge. I don't necessarily know how to solve it on my own. Given what we're about as an organisation, and our values and strengths, I know there must be lots of options. What does the team think?

Whenever there is a leader who truly embraces this approach leading a change project, the impact is incredible. Their confidence in what they are doing and why, and their clear sense of where they are ultimately headed, never waivers – and that is apparent to those around them throughout, and confers confidence and reassurance. But such leaders call it out, loud and clear, when it becomes apparent that a number of strands of work in pursuit of the ultimate goal are simply not working. And then they pull a team of people together to develop options and ideas and implement them. As a direct result of that team approach, the leader is able to generate more ideas, and better ideas, and implement them faster, and with a much greater level of engagement, than anything anyone would have been able to achieve alone.

In my experience, leaders who demonstrate some vulnerability in this way bring a multi-layered benefit to their organisation. First, and most obviously, they maximise the chances of finding a good solution to the particular problem in hand, by opening the floor to all of the talent, expertise and experience within the organisation. This is the coalface mastery point above.

Much more importantly, they establish the conditions which will make the generation of new ideas and the successful resolution of future problems much more likely. They are role-modelling the vulnerability and curiosity required to tackle complex issues. They are creating a space for others to step into and create and deliver. They are demonstrating a willingness to take risks and cultivating trust, while also being realistic.

If you 'get' vulnerability you understand that neither you as a leader, nor your organisation, nor your people, nor your clients need to be perfect. Organisations are on a perfectly imperfect journey. But if the management intent and ambition is always there, you simply keep working at it.

In my experience then, a certain vulnerability amongst the leadership team is a prerequisite if an organisation is to be free to take an evolutionary approach to setting strategy and managing change, without descending into chaos. In turn, it is the clear sense of purpose in an organisation that both encourages a willingness to be vulnerable (because it also engenders a deep-rooted sense of confidence, that we will consider in the next chapter), and enables it to deliver the right result – because the shared understanding of priorities, and the innate motivation to step in and help are prevalent across the organisation.

Put in place a framework within which vulnerability can happen, and the potential prize is enormous – trust, empowered and engaged teams, creativity, initiative, innovation, action. Collective intelligence. The framework within which vulnerability can happen most safely – and with the best chance of delivering what it promises – is a framework of purpose, values and stories.

One final thought on vulnerability as an essential ingredient in an evolutionary approach. There is not yet much research on what it might mean for organisations themselves to have the capacity to be vulnerable – systemically, culturally. At first blush, this sounds like anathema in a corporate world that is red in tooth and claw. Even those companies which are deeply values-led and well down the road of building authentic and open cultures internally and with their stakeholders would, I think, hesitate to admit to corporate vulnerability.

But what about organisations who are oriented towards recruiting other organisations to help them to solve problems? What about the impact of vulnerability when it comes to professional advisors or consultants working collaboratively with software developers, data analysts, regulators and industry bodies to help solve problems for clients? What about social services, care-givers, schools, medical professionals and businesses coming together in new ways to solve some of the most pressing societal needs of our time?[4] When we come in Part Four to consider some of the broader context around, and implications of, a purpose and narrative-led approach to doing business, an orientation towards vulnerability is a key part of it.

Hope

Brené Brown's perspective on vulnerability chimes with a separate body of research by C.R. Snyder, which Brown references, which is concerned with how hope is created and maintained.[5] For hope, don't immediately think of Sunday School and rainbows – or not exclusively; for Snyder, hope is not warm and squidgy, and it is more than simply an emotional expectation of a good outcome. It is a cognitive process – a particular way of thinking, or orientation – made up of what Snyder calls a trilogy of goals, pathways and agency:

- we have the ability to set realistic goals;

- we are able to figure out how to achieve those goals, including the ability to stay flexible and develop alternative routes; and

- we believe in ourselves.

In a corporate context, self-belief – Snyder's third criterion, and the only basis on which anyone would step into the space created by a leader's 'I don't know' – is rooted in purpose and stories: we know, based on our history, who we are, why this matters and what we are capable of. We have a collective confidence that we can achieve difficult things, solve complex problems and overcome hard challenges.

Similarly, the capacity for individuals and teams to think and act broadly in order to achieve goals (Snyder's second criterion for hope) – without descending into the mayhem of a thousand people all pursuing different great ideas – is greatly enhanced if they share a common sense of purpose and values, and a bank of stories that they can draw on by way of analogy – 'when we had a challenge like this before, this is how we solved it'.

It is only when leaders have confidence that their teams share a common purpose and values that they can allow space to be created and broad thinking and acting to happen. And yet it is absolutely imperative that they are able to create the space. It quickly becomes impossible to address every challenge centrally – any central team will become a bottleneck, and the inevitable limitations of their experience, skills and personality will limit creativity and mean that angles and opportunities are missed. This is true in the shorter-term problem-solving context, such as the CMS example above – and it is also true for organisations in their longer-term thinking and planning.

Margaret Heffernan talks in many of her books and lectures about the pressing need for collective leadership in today's environment.[6] She makes the distinction between the *complicated* systems of the past – which may have involved many details and steps, but which were ultimately linear and one-dimensional – and the *complex* systems of today, which are networked and interdependent, multi-faceted and beyond the mastery of one organising mind. Generating options, assessing them and then delivering solutions in a complex context, requires different minds to work together.

Stop, Look, Listen and Think

This brings us to the third key trait of evolutionary organisations. They have vulnerable leaders, and teams rich in hope – and they have learned how to stop, look around, listen to one another, and think together.

The CMS example above, again, contains all of these components. The leadership team called a hard stop, and convened that meeting, rather than allow matters to drift. The leadership team only knew there was an issue at all, and could sense that now was the time for a different approach, because they had been sticking close to the frontline, watching what was going on, and listening to people. You can't react in real time – you can't be an evolutionary – from an ivory tower; you have to be in amongst it. The leadership team at CMS figured out *how* to respond by sitting together as a team, thinking through the options and, again, listening to one another.

Let's take stop, look, listen and think in turn. We saw earlier that one of the advantages of adopting an evolutionary approach is that it enables you to go fast. Much like freewheeling downhill on a bike, however, this is only an advantage if you also know how and when to stop. Years ago, when I was new to a job that was more about oversight and leadership and less about operational doing-the-doing, I used to have a recurring dream about obsessively monitoring a dashboard of dials and knobs and lights, scared to blink. Whenever one of the dials sank into the red zone, or a light flashed, I had to act, and fast. That's what knowing when to stop means: letting most things happen, but calling a halt either before trouble strikes, or in time to change direction in order to secure an even better outcome. It is essentially about prioritising, and its foundation, once again, is that really clear sense of purpose; coupled with a good nose for trouble, refined on the frontline.

Looking and listening are about closely observing what is going on around you, without having pre-judged the outcome. Asking questions, getting across the business talking to a broad range of people with different roles and perspectives, keeping an eye on KPIs. CMS had the various sounding boards and think tanks and email addresses that I have mentioned. They also used Crowdicity, a crowd-sourcing platform, to hear feedback and test ideas. But they also just tried to stay really close to people in how we did everything day-to-day. Physically sitting on the operational floors, travelling to all the offices, giving people a call, dropping them an email. Generally trying to show up as one human being, interested in another, and listening to their stories.

Thinking together is about truly collaborating to design a solution. Margaret Heffernan again, in *Beyond Measure*,[7] talks about a study by Thomas Malone and team of the Centre for Collective Intelligence at the Sloan School of Management at MIT in which the researchers studied groups who had proved to be particularly effective at creative problem solving, with the aim of identifying the salient features that made these groups so effective. The researchers found three key features which mattered more than the innate skills or intelligence of any individual group member. The first of these was that the members of the group gave one another roughly the same amount of time to talk. Although the groups did not formally monitor or regulate this, the upshot was that people contributed equally.[8]

Here's Heffernan again, this time talking about her own team in the light of the MIT research:

> We were all smart enough and had a wealth of different experiences, but no one deferred to anyone; that made us curious about what each could offer. We knew we needed an answer but we also knew that no one of us had it; we would have to work together to craft something we could not make alone. At times we were frustrated, scratchy, impatient. But nobody had any agenda. We all cared passionately about our shared success.

There's purpose showing up again, right in that last sentence. This is skilled listening and thinking, underpinned by a passionate commitment to come up with an answer. An approach rooted in purpose makes it much easier to try, test and embrace new ideas with confidence that one is not, as a result, being pulled off course. Confidence is what the next chapter is about.

Notes

1. attr. Picasso, Pablo in the exhibition 'Picasso 1932 – Love, Fame, Tragedy' at the Tate Modern, London, March 2018.
2. Ariely, Dan, *Payoff: The Hidden Logic That Shapes Our Motivations*, Ted Books, Simon & Schuster, 2016 and Ariely, Dan, 2010 , *What Makes Us Feel Good About Our Work*, video, TEDx, viewed April 2018.
3. Brown, Brené. *Daring Greatly – How the Courage to be Vulnerable Transforms the Way We Live, Love, Parent and Lead*, Penguin, 2012.
4. As envisaged in Hilary Cottam's Radical Help (Virago, 2018) – a compelling manifesto for a re-imagining of the welfare state.
5. Snyder, C.R., 'Hope Theory: Rainbows in the Mind', *Psychological Inquiry*, 2009.
6. For example, Heffernan, Margaret, Forget the pecking order at work, Ted-Women 2015, viewed 11 April 2018.
7. Heffernan, Margaret, *Beyond Measure: The Big Impact Of Small Changes*, TED Books, Simon & Schuster, 2015.
8. The second and third criteria are revealed in the chapter on Understanding – keep reading (or read Heffernan!).

Chapter 11

Confidence

We can still astonish the gods in humanity
And be the stuff of future legends
If we but dare to be real,
And have the courage to see
That this is the time to dream
The best dream of them all.

Ben Okri[1]

How come pretty much every corporate buzzword begins with a C? Courage, conviction, culture, creativity, curiosity, communication, cooperation, collaboration, conflict, client, customer, coaching, celebration, champagne, cake, cheese . . . oh, wait.

Certainly there were a lot of contenders for the title of this chapter. But 'confidence' it is, partly because it seems to me the most striking of the various concepts in terms of its importance in a change context, and partly because it encompasses so many of the other c-ideas listed above. (Also, a chapter on cheese would have been odd.)

In this chapter, we will explore what confidence means – and what it does not mean – and consider how it manifests itself in organisations, and what advantages it confers. I will then outline the various ways in which periods of change can present a challenge to confidence and how organisations can nurture and grow confidence, including by using the foundational concepts of purpose and story.

What Is Confidence?

Let's start with what confidence is not. Confidence is not dogmatism, or arrogance. It is not foolhardy, or superior. It does not swagger. It is not relentless Pollyanna-ish positivity in the face of all evidence to the contrary.

The word 'confidence' itself comes from the Latin *con fides* – meaning with trust or with faith. In yourself, in others, in circumstances. By definition, confidence does not equate to certainty. It has a kind of knowing-unknowing right at its heart. Confidence is about stepping into a gap; it is about participation and commitment – throwing your hat in the ring. Confidence comes from experience and it is wholehearted and laced with humility. Confidence says: 'I don't know for sure that I am right, and nor am I sure that this path is the right one, but I have listened and debated and reflected, drawing on all my experience, and now I am choosing this way forward and will do everything I can to make this work.'

Confidence, then, does not exist in isolation. It is a posture in relation to ourselves and others. We can have confidence vis-à-vis our team and colleagues, our clients, or market, and/or our competitors. We can think about confidence at the level of the individual, and for leaders and teams. We can also think about confidence at the corporate level – being a confident organisation. In this chapter, we will explore all of these aspects of confidence.

How Does Confidence Manifest Itself?

There are a number of striking ways in which confidence shows up in individuals and in organisations. Throughout the CMS merger, I watched the senior leaders role-model and exude confidence in a number of ways. I saw them, for example, being always available to listen to people – their concerns, ideas and opinions, both informally, and in thrice-weekly 'listening' coffee sessions. Without fail, those sessions would yield a new idea, which the leaders would then set in motion. I saw clarity and decisiveness in relation to the myriad decisions, big and small, which came across the leadership team's desk – never flinching at the unexpected, dealing with politically fraught decisions around roles, conflicts, whatever, with clarity and astuteness, and always prioritising the person and the relationship at the heart of any issue. The leaders also retained a twinkle in their eye throughout the whole period, never forgetting to stay curious, play a little, enjoy the sheer privilege and challenge of what we were engaged in.

If we unpack some of what I have just described in the CMS approach, we will find some of the key aspects of how confident people and organisations tend to show up. Confident people and organisations:

- **transcend and include** – celebrating, embracing and incorporating a wide variety of ideas, opinions and ways of doing things, with a view to achieving something even better;

- **approach challenges and opportunities** *con brio* – making bold, swift, principled decisions;

- **deal with conflict** well;

- **hold their nerve** when things do not go quite to plan or take longer than expected to come to fruition; and

- **play** – staying curious and trying new things.

Let's consider each of these in turn.

Transcend and Include

I first encountered this phrase in the work of the writer and theologian Rob Bell,[2] who may in turn have borrowed it from psychologist Ken Wilber.[3] They both use it in the context of spiritual and psychological growth to express the notion that as we grow and develop we can do so in a way that involves accepting and embracing our former ideas and identities, rather than ditching and disowning them. It makes me think, too, of the golden rule of improvisation: whatever curveball someone throws into the mix, a fellow actor must never reject it out of hand, but rather must respond with 'Yes, and . . .' and find a way to continue the narrative.

In the context of the CMS merger, I used the concept of transcend and include as a way of thinking and talking about that 'fourth way' which kept the best of how the three legacy firms did things but also sought something new and even better for the combined firm. This involved preserving and celebrating some important aspects of each firm's past. It also involved being profoundly respectful of history and peoples' sense of identity, and giving people time to grieve what they were moving away from and become accustomed to what they were moving towards. And it involved being ambitious and committed to creating something that embraced all of that and yet was even better.

This kind of mindset is open and inclusive, not fixed or defensive. As such, it depends on, and demonstrates, a degree of confidence on the part of individuals and leaders.

Con Brio

A second marker of confidence in organisations is an approach to decision making, problem solving and execution that always makes me think of the musical term *con brio* – with spirit, or with vigour. A bouncing conductor keeping time, frantic bowing of violins, a cymbal crash, enter trumpets . . . Confident organisations, and confident people and leaders within them, make good decisions quickly, but in a coordinated way and based on principles and purpose rather than (only) reacting to circumstances.

Conflict

While working on this chapter, I had a disagreement – a trivial one, over something inconsequential which was quickly resolved – with a close colleague.[4] At the time of the argument, I was rushed and tired and not at my most confident. As a result, I reacted badly. What does 'badly' look like? Well, internally, I took the criticism personally, and believed it to be a comment on my ability, judgment, integrity and general worth as a human being. We explore more below what was going on for me at that point. Externally, I defended myself, and attacked my colleague – telling him he was wrong and, actually, a little bit annoying.

Had I been feeling more confident, what might I have done instead? Well, internally, I might have recognised and accepted my colleague's input for what is was: namely, useful information, wrapped in opinion. I would have been able to hold it, reflect and consider it in the light of any other feedback received and my own observations. Externally, I could have thanked my colleague for his input, presented an alternative argument and/or acknowledged the truth in his, talked about a practical way forward to improve things, and maybe even made a bit of a joke to diffuse any bad feeling . . .

At a firm level, a confident approach looks very similar. It is confident organisations who hold their hands up in the market when they make a mis-step, who overcompensate customers for poor service, who actively seek feedback on their service, who actively enjoy the cut and thrust of market competition and engage in ways that support and strengthen their brand and values, and who surprise, disarm and delight their customers.

Holding Your Nerve

At times of change and disruption, there will inevitably be things that go wrong, or don't go according to plan, or take longer than expected. If confidence is low (and, as discussed below, there are plenty of reasons why this is likely to be the case during such a period), delays or deviations of this kind can knock it further. Worse still, if you're not feeling confident on one front, it's easy to lose your footing on other fronts too, like scrambling up a scree slope. In the face of a direct challenge, obstacles, doubt or someone strident with another idea, it can be tempting to stop, or change course, or hitch one's wagon to something new.

Confident people, and confident organisations, however, resist this. We see them facing down market uncertainty, criticism, slow uptake of new ideas, competitors . . . and remaining clear and strong about their purpose, focus, priorities and direction. Often, the early days of a change project are an exercise in holding one's nerve; keeping connected to one's original purpose in having embarked on this project, trusting the people and processes one has in place, and then waiting it out.

As we have considered, being confident does not mean being relentlessly positive. But positivity is an important tool for leaders when working to maintain and grow individual and collective confidence in organisations. Research shows that we do generally trust and follow those who face challenges with positivity and poise. This is called the confidence heuristic – our brain basically takes a shortcut and says, 'okay, if this woman looks and sounds as though she knows what's what, she probably does know what's what'.[5] Of course leaders need to strike a balance, and acknowledge issues, but instilling confidence is a key part of their role.

Playfulness

One of my more extrovert friends, faced yet again with some cavilling from me about networking, once advised me to face daunting social events as my 'one-glass-of-fizz' self – by which she meant with the very slight disinhibition and elevated sense of fun that comes inside a glass of champagne (but without actually getting to drink the champagne). Think of this best, energised, grounded version of yourself. Perhaps you laugh more, muck around with your kids, sing to yourself, flirt, throw a few more sticks for the dog – all just for the fun of it. This is playfulness, and it only really catches, only ignites, when we are feeling confident in ourselves and in our environment. In confident organisations, this playfulness shows up internally in high levels of engagement, celebration and rituals, in a willingness to try new things, think differently, and welcome challenge and debate. Externally, it is about taking new ideas to market, experimenting with ways of engaging with customers, investing discretionary effort, 'giving back' through CR, driving thought leadership, participating in the sector.

Sounds Great, How Can We Get Some (and Why Is It So Hard)?

Throughout my career, the single area others – mentees, coaching clients, colleagues, friends – have most often sought out my advice on is confidence: not having it, wondering why they don't have it, thinking everyone else has more of it than they have, wishing they had it and – above all – wanting to know how to get it. Confidence, then, can be an elusive thing at the best of times. On top of this, periods of profound change present some particular challenges to our ability to feel confident.

In my experience, there is no single failsafe method for acquiring confidence. That said, over time, I have seen individuals benefit from focusing on four areas in particular, and it has been interesting to think about how these areas also apply at an organisational level. It is also interesting to me to note the marked extent to which the language and metaphors in relation to each of these areas is so rooted in the earth and the ground; it seems that we experience confidence in a very visceral, physical way.

The four areas of focus then, are:

- reframe
- build your cairn and stand on it
- stay rooted
- take your space

Let's consider each of these in turn as they apply to individuals, leaders and whole organisations.

Reframe

The first thing that helps with confidence sounds a bit like the oldest trick in the book – 'fake it till you make it'. But this is about tricking yourself, rather than anyone else – and it's not so much about tricking, really, as persuading . . . This is about telling your brain a different story: 'This is not an argument; we're exploring ideas'. Or, 'I'm not nervous; I'm excited.' This is called cognitive reframing.[6] The aim is to keep the brain from perceiving a threat and so retreating into fight or flight mode, at which point our capacity to think clever thoughts reduces dramatically. Think about Rock's SCARF model again – if we can preserve our status, increase certainty slightly, and frame matters such that we can see that we have a degree of autonomy, we will be better able to stay playful, curious and resourceful, rather than succumbing to defensiveness and anxiety.

Leaders can do a lot to inculcate and nurture this sort of capacity to reframe in their people and teams, in particular by role-modelling, being deliberate in the language they use – speaking of opportunities and excitement, rather than threats and concerns, for example – and by telling stories to showcase success.

The corporate equivalent to this cognitive reframing at individual and team level is also around thinking in terms of opportunities rather than threats; moving towards rather than moving away.

This is not glib, and it is not simply a question of language. It is an intentional choice, made time after time after time, to adopt a particular posture vis-à-vis the world. Practised regularly, it will influence our culture, our values and our brains - keeping us engaged and resourceful, and building our confidence in ourselves. There is an important interplay between the organisation and the individual here – a confident climate in the organisation will help to reduce individual anxiety during an ongoing period of uncertainty, and so make it easier for individuals to remain out of threat state, and to keep activated the part of the pre-frontal cortex that helps them to regulate emotion and make good decisions.[7]

Build Your Cairn, and Stand on It

When I was growing up, my mum and dad would pile my sister, brother and me into our clapped-out old Sierra Estate every summer weekend, and head north from Edinburgh to go Munro-bagging in the Scottish Highlands. A Munro is a mountain over 3000ft in height – there were 284 of them at the last count – and to 'bag' one is simply to climb it. Depending on the weather, which cassette had been played in the car, and how many Mars Bars there were in the rucksack, this activity elicited delight, tolerance or – most usually – grumpiness from me and my siblings. However deep our sulks, though, one thing all three of us always enjoyed was looking out on the ascent for a suitable stone that we could slip into our pockets to add to the cairn when we finally got to the top.

A cairn is a human-made bundle of stones, used for a variety of purposes since prehistoric times, and most commonly these days as a landmark to mark the summit of a hill, or as a trail marker. Once we had chosen our stones, we would each clutch them in our pockets and then, at long last and despite having been apparently close to dying of boredom just a few moments previously, race the final few metres to the summit to place them into the bundle.

Now. There is a surprising amount of controversy and etiquette surrounding all sorts of aspects of cairns in hillwalking circles, but the general consensus is that, having climbed the hill, you shouldn't really also climb the cairn itself, presumably because you risk dislodging the stones carefully placed by the many travellers who have walked this way before you. But that never stopped my little brother. We have several pictures of him, at around seven years old, all scrawny legs, cagoule and curls, standing atop a cairn, shrouded in mist, fists raised in triumph, yelling his sheer joy at being on top of the world.

In my experience, one of the most effective ways of building confidence is to take some time to intentionally pile together your successes and achievements like pebbles in a cairn. Each of them might seem small and insignificant, but together they make a pretty substantial pile. Few of us are very good at doing this. Busy, high-achieving people have 'to do' lists galore, but how many of us convert our 'to do' list to a 'done' list? We go too fast, we don't stop to reflect, and so we underestimate, discount or just plain forget our achievements. But if we want to build confidence, we should develop the habit of noting those everyday successes; of gathering together those small pebble stories that may not look like much on their own, but together build a cairn. We should tell ourselves those stories; rub them under our thumb, breathe on them and polish them, and remember what we did. We should pile them all up together.

And then, like my little brother, we should scramble up that cairn, stand on it and yell – because it also helps us to build confidence if we can find ways of credibly and authentically, and without boastfulness or arrogance, talking about our achievements and presenting them to others. In *How To Have A Good Day*,[8] Caroline Webb describes the practice of a sales executive who blocks out half an hour every Friday afternoon to note down her achievements for the week – gathering the pebbles – and then to figure out who might be most interested in hearing about each example. By sharing small snippets of relevant and helpful information about her achievements with others, this woman describes herself as having built a personal brand and broadened her network with people who actively respect and endorse her.

Again, this lesson applies to us as individuals, as leaders of teams, and to organisations as a whole. One of the challenges we had post-merger was that individuals worried that their track record was somehow lost; their sense of standing and goodwill that had taken them a whole career to build. This worried people insofar as they felt they depended on their reputation to enable them to earn respect in the eyes of colleagues and clients, but it also went to their own self-confidence. Individual cairn-building can help, and so too can deep respect and intentional encouragement from leaders; curiosity about what individuals have done in the past, celebration and marking of achievements in the new firm. At a neurological level, reminding people of past achievements, or asking them to talk about something they did well or something that makes them proud, can generate reward-related dopamine, just as if they were experiencing that achievement again right now.[9]

Leaders can also recognise and celebrate the achievements of the team as a whole, and of the firm. At CMS immediately post merger, the latter meant drawing a lot on the legacy firms' respective histories – *we are the product of three great firms with great histories* – and this helped the leadership team both to demonstrate real respect for the legacy firms and past achievements, but also to remind people how capable they are.[10] Very quickly, though, the leadership team was able to begin collecting and bundling together the pebbles of achievement in the new firm – pitches won, new panel appointments, recognition in league tables, rewards, directories, the press . . . They kept a list and used stories of successes large and small whenever the opportunity presented itself. Going back to the confidence heuristic, telling the new firm's stories, really made a profound difference to its own sense of confidence as a firm, and to other stakeholders' confidence in the firm – clients, the market and the press all taking note that here is a firm that seems to know what it's about and that means business.

One final thought, just to pile a last pebble on this teetering cairn metaphor: while researching the history and etiquette around building, adding to and climbing on cairns, I came across a lovely blog,[11] in which the author pointed out that cairns are often built by someone at the time at which they are feeling at their least confident:

Typically, a person building cairns is inexperienced with the area . . . If I place a cairn it's because I am unsure I will notice a spot on return . . . More often than not, cairns are placed on the way up, not on the way down.

I love that! It's precisely when we are least sure of the road ahead, least confident in our own wayfinding skills, most vulnerable (that word again!) that we should take the time to collect those pebbles and build that cairn.

Stay Rooted in Your Purpose and Values

There's an infamous test called the Trier Social Stress Test in which subjects have to give a short speech to a panel to explain why they are the best person for a dream job. It is an ordeal deliberately designed to induce stress: people are put on the spot in various ways and the judges are not allowed to smile at or encourage the subjects in any way. In a UCLA study,[12] a group of volunteers who were about to undergo the Trier test were asked to rank a list of values that were important to them. A sub-group were then asked further questions to elicit their feelings about the values they had ranked as most important to them. The study found that this sub-group – the group that had tapped into their feelings about their values – both felt less stressed by the Trier test that followed, and had lower levels of cortisol in their saliva.

This is good supporting evidence for what I have seen work well in practice – namely, that people who manage to stay close to what matters most to them during periods of uncertainty or change or stress, seem both to retain a deeper-rooted sense of self-confidence and to convey a sense of calm and confidence to those around them. Leaders who can do this are particularly effective and valuable at times of change because, as we have seen, that outward confidence is contagious. By extension, organisations that know how to tap into their purpose and then convey it will also acquire and inspire deeper confidence in their people and other stakeholders.

Webb suggests three practical steps to help people to tap into their values before stepping up to a challenge, namely:

- to write a couple of sentences about their broader aspirations in life;

- to remind themselves of the noble reason behind the specific thing they are about to do; and

- to focus on whatever they feel most strongly about.

Purpose and stories again. Writing things down brings clarity and a stronger, more concrete narrative. It also taps into the logical part of your brain.[13] On top of this, Webb's approach has people tapping directly into various layers of purpose and values – into one's own overarching purpose, the task-specific purpose, and one's values around the wider context in which the task is happening. It is interesting how specific and intentional this all is – deliberate attitudes and practices and resultant actions, rather than a vague and fuzzy sense of moral good.

Again, it is easy to see the read across into leadership and on into the broader organisational context. We need leaders who can paint the noble, big picture backdrop to the quotidian task at hand, who can talk about the aspirations and purpose of their team and organisation, and help people to focus on what the group feels most strongly about, amidst endless distractions.

Take Your Space

The final strategy for building confidence to explore is another very physical one, with a couple of different aspects to it, around 'taking your space'. In this context, we can think about both physical space and temporal space.

Physical Space

Being an introvert, a woman and a relatively young person in the various teams I worked in, for a long time in my career one of the things I struggled with most was how to make a contribution in meetings when it feels like a jungle – lots of men chest-beating and waiting (or not waiting) to speak, nobody listening or deferring. I would have a point sitting on my tongue like a hot pebble – often a pertinent fact that people actually needed in order to help them make a better decision – and I would wait and wait and wait, and then the moment would pass and I would have to swallow it, and it would sit there for the rest of the meeting, burning like shame in my chest. Or, almost worse, I would mistime it, spit the wretched point out, and it would either be missed entirely and sit there sizzling on the table, or it would be picked up by some big beast who would then bide his time and then pass it off as his own pearl of wisdom. Argh.

When I got my first big role which was going to involve being in the thick of this jungle almost constantly, a very wise older female mentor gave me three very specific pieces of advice, which count amongst the most valuable I've ever received, and which, fully ten years later, I still pass on to someone else almost weekly.

The first is never to sit in a seat at the end of one of the long sides of a board table. Those are the least powerful seats – it is hardest to interject from there, hardest to make eye contact, and hardest to project your voice.[14] The second is that when I want to speak, I should shift my weight all the way forward in my chair, place my elbows on the table, and put my hands in a steeple position, fingertips lightly touching – a classic power move in body language terms.[15] The third piece of advice is a mind trick, which I suspect has its origins in NLP, which is to visualise, on entering a room, or taking a place at the table, the physical 'personal space' which you currently occupy and then (eyes closed and deep breathing if possible) to imagine doubling the diameter of that space. And then doubling it again. And again. When I manage to do this well, it feels like ripples of power and connectedness radiating out across a pond.

The apparently magical power of these three pieces of advice is well supported by research conducted by Amy Cuddy and her team and reported in her wonderful book *Presence*.[16] The ground-breaking finding is that the mechanism which leads us, in common with other primates, to adopt large, open postures when we feel confident, works in reverse too – which is to say, that by adopting large, open postures, we increase our sense of confidence. Cuddy advocates adopting 'power poses' (think Wonder Woman) in advance of a big challenge. I am also thinking about my sporadic yoga practice, and about how much more confident I feel on a day when I manage to squeeze in even just a couple of sun salutations.

So far, so good, but what is the read-across of all of this very personal confidence-building for organisations? Reflecting on the lessons and underlying themes, I think, in a nutshell, the key messages for organisations are around:

- adopting the right posture (power poses, imagining big personal space);

- being in the right position (avoid the corner seat!); and

- being ready to convey your message (weight forward, fingers steepled).

CMS had the first two of these absolutely nailed – it knew the kind of all-new, future-facing, purpose-led firm it wanted to be, and had the self-belief and culture it needed to get there. The merger was a transformational move to position the new firm in the seat of power at the table, and all the work the leadership team did, internally and externally, to find and tell the stories about the firm, ensured that they were ready to speak, and able to make their voice heard. This enabled the new firm to focus on conveying its message. The firm came to understand that taking its space in this way too – in airtime, column inches and trophies – is an important part of building confidence. Communicating can also be a bold statement of intent: it has a galvanising effect and can endow the organisation with the extra shot of persistence and determination that it needs to follow through.

Temporal Space

The second kind of space we can think about taking is space in time. Just as we tend to adopt postures that shrink our physical selves when we are feeling uncertain or unconfident, we also tend to try to minimise the time we take on things – particularly when it comes to sharing our presenting our own ideas. We rush. But slowing down and taking the time we need can help us both to build and to convey confidence, as individuals and as organisations.

If the advice from my mentor about how to sit at the board table was game-changing in terms of helping me to manage my physical presence, the advice I received from my former coach and dear friend, the late Juan Coto, transformed my ability to build confidence by taking time. Juan's advice was very much around building or maintaining confidence in the particular moment of acute stress, but the lessons apply more widely. Juan was a coach to elite athletes, particularly tennis players, and the process he developed depended on bringing the focus right into the four squares of space right at the centre of a (in my case imaginary!) tennis racquet.

With Juan's help, I came to understand the emotional wheel that would start with nervousness and self-doubt, move into 'What ifs . . .' , throw off a number of horrible physical symptoms (clammy palms, dry mouth, dodgy tummy), and then likely lead to an actual error in performance. This in turn would lead to anger, guilt, shame, self-recrimination, which in turn would lead to an 'I always do x . . .' type of self-sabotage. Round and round it spins, faster and faster, bigger and bigger, like a Catherine Wheel throwing off sparks. But Juan would say, 'Stop!' and urge his coachees to zoom their focus right in to those four tiny squares, representing four things to do. The first is to breathe, slowly and deeply. The second is to treat yourself to some kind self-talk – talk to yourself as you would talk to your best friend. The third is to visualise exactly the outcome you want as precisely and with as much detail as you can, and the fourth is to identify the smallest next step to get there – i.e. 'I can achieve this, as long as I . . .' . In the heat of the moment, all of this takes a matter of seconds – the time it takes to walk back on court to take the next serve – but it is enough to restore equilibrium and rebuild confidence.

In an organisational context, the 'holding your nerve' aspect of confidence depends on taking time. This is not about being indecisive, but it is about allowing things to evolve, as we learned in the previous chapter, and about taking a longer-term view when it comes to assessing performance and priorities.

Taking time in an organisation is also about celebration and ceremony. It is about celebrating large and small successes, and telling stories about them; weaving them into the fabric of the firm. It is about developing rituals ('We always have drinks on a Friday', 'We always have sweets on payday', 'We always have a children's Christmas party'). It is also about allowing time for grieving the passing of things – particularly on merger – and about respectfully marking milestones. There is something about marking the turning of the seasons and the year with routines and rituals and celebrations that really seems to build engagement and confidence, internally and externally – it makes an organisation feel like an establishment, a permanent fixture.

One final thought on this idea of 'taking your space' is around the proactivity implied by that word 'take'. Not 'wait until a space presents itself and then quietly sidle into it', but 'take it!' This itself is an act both born of confidence and liable to create it. The realisation as an individual, team or organisation that you are in charge, that you have agency and can influence the outcome, is empowering and confidence-building, but it can be daunting to take that first step. That's what makes Juan Coto's, Amy Cuddy's and Caroline Webb's advice in this area so valuable – it is all broken down into practical proactive steps that individuals and organisations can take. That said, all of their advice is based on research which also points to the deeper need to be firmly rooted – to understand your purpose and have a story to tell about it. This is where true confidence comes from.

Notes

1. Okri, Ben, *Mental Fight*, Rider, 2012.
2. e.g. Rob Bell, quoted in article by Jo Dolby, 'What I Learned From Two Days With Rob Bell', *Christianity Today*, 2 April 2015.
3. e.g. Ken Wilber, referenced in article by Pearson, 'Transcend And Include: The Integral Attitude Towards Competing Perspectives', 29 January 2015 https://philosophadam.wordpress.com.
4. My daughter calls these 'collifriends' – an accidental portmanteau which I love, but which always makes me think of wagging tails and big doleful brown eyes.
5. Zarnoth, P., and Sniezek, J.A., 'The Social Influence of Confidence in Group Decision Making', *Journal of Experimental Psychology*, 33(4), 1997.
6. In cognitive therapy, the term more often used is 'cognitive restructuring'. This is a formal psychotherapeutic process of learning to identify, challenge and so change irrational or maladaptive thoughts, developed by Aaron T. Beck in the 1960s as a key component of cognitive therapy. I am not a therapist, and so no expert, but in general terms, while restructuring refers to the specific form of reframing that is done, supported by therapy, in a structured way to support positive changes to thinking patterns, reframing is a broader concept and can be positive or negative; conscious or unconscious (as with the rose-tinted spectacles of hindsight). I am using the broader term here advisedly in order to connote more minor changes which people can generally achieve without the support of therapy.
7. Scarlett, Hilary, *Neuroscience for Organizational Change: An Evidence-Based Practical Guide to Managing Change*, Kogan Page, 2016.
8. Webb, Caroline. *How To Have A Good Day*, Macmillan, 2016.
9. Scarlett, Hilary, *Neuroscience for Organizational Change: An Evidence-Based Practical Guide to Managing Change*, Kogan Page, 2016.
10. Bridges, William, *Managing Transitions: Making The Most Of Change*, 4th edn, Nicholas Brealey Publishing, 2017.
11. www.willhiteweb.com/info/cairns/what_are_cairns_426.htm.
12. Creswell, J.D., Welch, W.T., Taylor, S.E., Lucas, D.K., Gruenewald, T.L., Mann, T., 'Affirmation Of Personal Values Buffers Neuroendocrine And Psychological Stress Responses', *Psychological Science*, 16(11): 846, 2005.
13. Research in the context of goal setting, by Professor Gail Andrews, Dominican University, California, presented in May 2015 at the Ninth Annual Conference of the Psychology Research Unit of Athens Institute for Education and Research, and summarised at www.dominican.edu/dominicannews/study-demonstrates-that-writing-goals-enhances-goal-achievement.

14. So often, in all sorts of organisations, when I walk into a board meeting as a guest, all four of those seats are occupied by women. It's interesting. I have before now texted all four of my corner-dwelling sisters in the course of a meeting and suggested that they change seats when we next have a loo break. . . .
15. Even now, I'm rarely alpha enough to have a fully raised steeple, as it were, but I find a horizontal steeple or even a lowered one (fingers pointing down) works wonders.
16. Cuddy, Amy, *Presence: Bringing Your Boldest Self To Your Biggest Challenges*, Orion Publishing, 2016.

Chapter 12

Agility

We shape ourself
to fit this world
and by the world
are shaped again
David Whyte[1]

If there were a hundred possible contenders for the title of the last chapter, there were only ever two A's in the running for this one. I wanted to call this chapter 'Agility' because I wanted it to be the place where we could explore what it means to be able to move, as organisations and as individuals, quickly, nimbly and elegantly into a new space. This is different, although linked, to the idea laid out in Chapter 10 that organisations should strive to be evolutionary in their approach. Agility is about a leap more than a sideways shuffle, and the impetus for it tends more often to be external – a shift in the market, the tantalising glimpse of a new opportunity as it dances past.

But Mark Humphries, the former CEO of Huntswood, who led its cultural transformation and subsequent rebrand in 2005, and is now a non-executive advisor to several lucky boards and, it turns out, to me, suggested that this chapter should be called 'Acceleration' instead. As we talked through the key ideas in this book over coffee, testing them against his experiences at Huntswood, we got to talking about purpose and momentum. Mark says that, in his experience, a sense of momentum matters almost as much as a purpose, which is to say that while some people get on the bus because they are excited about where the bus is going, some people get on the bus because they are excited just to be moving. I love this idea, and the pictures it conjures for me of Sunday School outings in Scotland when I was a little girl – all of us squeezed on the bus with our sandwiches, brightly-coloured paper streamers hanging out of the windows, everyone singing. Or of all-night Greyhound trips across Australia – the whole world asleep and the impossibly wide sky aching with stars and possibilities.

I agree that this yearning, this desire just to move, is critical, and goes to the heart of what it means to lead or participate in organisational change. I also understand that it's easier once you have the wheels rolling, to keep them rolling – this idea of momentum, too, is fundamental. But 'Acceleration' also made me think of that bus getting faster and faster

– careering off, bouncing, down a hill. There is a lot of talk about the pace of change accelerating, and the exponential increases in the amount of information available to us, processing speeds, and so on are truly staggering. But that does not lead inexorably to the conclusion that we must all run faster and faster in a vain attempt to keep up. Rather, we need to find a way to step back a little, reflect and then take the opportunities which are best for us.

So. Agility it is.

What Is Agility?

I am truly terrible at gymnastics. A simple cartwheel is beyond me. I am pushing my luck with a wobbly handstand. All limbs and lumps, as a child I would sit with my knees curled up under me on the sofa, reading, while my sister perfected her crabs, bridges and backflips on the living room floor and gradually won badges which we sewed up the length of her leotard. 'I'm better at gymnastics than my sister,' she would tell everyone, 'I'm better, because . . . I'm FRAGILE.'

Picture a gymnast. Or a ballet dancer. Or one of those incredible breakdancers you see body-popping a metre off the ground in any European city square. All of them able to bend their body into incredible shapes – to spin and leap and draw a perfect arc through the air. All of them bendy beyond belief, and incredibly powerful. This power comes in large part from their backs – the strength in their spines, in particular, is remarkable. They are agile, not fragile. Agile, not fragile. We should get that on a t-shirt. Little sisters take note.

A McKinsey report published in January 2018[2] describes an agile organisation in terms of its operating model, talking about a structure that is designed for both stability and dynamism (agile, not fragile!) which is flat in terms of hierarchy, highly networked, and people-centred. An agile organisation, says McKinsey:

> operates in rapid learning and fast decision cycles which are enabled by technology, and is guided by a powerful common purpose to co-create value for all stakeholders

I Don't Dance

I am working hard in this book to avoid buzzword bingo, and Agility perhaps sounds a little bit suspect. Is this just an of-the-moment concept, borrowed from tech, without much substance or business case behind it? Will it cartwheel itself off to someplace else if I just stay curled in my reading corner a little longer?

No. The important point to grasp here is that a tendency towards greater agility is both a natural consequence of, and an imperative driven by, some profound shifts which are happening in organisational power structures *in any event*. The traditional paradigm, conceived in the industrial age, is static, hierarchical, patriarchal and neatly organised into silos and tiers. Power and information are owned at the top, and goals and decisions flow down through the hierarchy. This is a strong and stable structure, but it is also rigid, slow and stiff – it could no more do a backflip should the need arise than I can.

It is becoming increasingly obvious that this structure is not fit for purpose in our new 'digital age'. Organisations operating within this paradigm struggle to keep pace with a quickly evolving environment in which customers, investors, competitors, regulators and collaborators are all highly demanding, and their demands are constantly changing. They struggle to adapt to the constant introduction of disruptive technology. They are ill-suited to the exponential increase in the volume of information, its democratisation, and the resulting need to communicate in every direction all the time. And they are wrong-footed in the new war for talent, which plays out on a flat field and favours the shape-shifting, autonomy-granting, learning-bestowing new generation of organisations.

So what we see happening now is that the traditional models which have served us well for centuries are evolving into much flatter structures with a more collaborative and collective approach to leadership.[3] The aim is to retain the stability which the traditional models have, but also to introduce a much more dynamic element. The go-to metaphor in this context is that strong gymnasts' spine: the most agile organisations build a strong backbone of core elements – e.g. purpose, structure, governance, customer base, processes – that can then support a range of dynamic capabilities, enabling the organisation to get going, empower people and act quickly to meet new challenges and embrace new opportunities. Netflix takes a similar approach, which they famously describe as being 'highly aligned, loosely coupled' – meaning that they are very clear on their over-arching purpose and strategy, and have a structure that supports these – but within that they operate on trust, and allow teams a high level of autonomy in terms of tactics and execution.

Our reason for discussing Agility here, now, in this book, is slightly different to our reasons for having discussed Belonging, Evolution and Confidence. During periods of change, belonging and confidence are both challenged and we *need* to work on them if change is to be successful. When you are in a period of change, adopting an evolutionary approach works better than being overly planned and dogmatic. But the point about Agility is, first, that the shifts described above further strengthen the case for, and provide a further catalyst for, change; and secondly, that if we are changing anyway, we might as well take the opportunity to do everything we can to hardwire the key features of agility into our organisation.

Okay, Show Me Some Moves

So, what does agility look like? McKinsey identifies five trademarks of agile organisations:

- a 'North Star' embedded and embodied across the organisation – that is, a shared purpose and vision;

- dense networks of aligned, empowered and accountable teams, balancing coordination with individual autonomy;

- rapid decision and learning cycles, based on experimentation, performance, transparency, and standardised ways of working;

- a dynamic people model which engages and empowers people, fosters a strong community and builds the skills for innovation and agility; and

- enabling technology as an integral part of how they do business.

Together, these five trademarks, says McKinsey, enable an organisation to achieve a good balance between stability and dynamism and to take advantage of the many opportunities that an uncertain and volatile world presents.

Okay, so far so good, but this is a seriously hardcore list. As a set of descriptors for those organisations that are already role models for agility, it is impressive, but how does an organisation that would like to get more agile go about building capability in relation to each of these five criteria? And where and how might purpose and stories help us?

Well, taking them in turn, the first of these five speaks for itself, and is all about alignment, which in turn is a pre-condition for agility. As discussed in Chapter 10 on Evolution, if everyone understands and is bought into a common purpose, then all sorts of coalface action and decision making can happen quickly and easily. Stories and storytelling can help with the discovery, development and reinforcing of a common purpose, and purpose builds alignment.

The second criterion is about how people build relationships, communicate, share ideas, collaborate and balance autonomy with holding each other to account. The third is about how decisions are made and how people learn, and the fourth is around the system within which people are asked to operate and how well it supports them to work in an agile way. These are complex and fascinating aspects, where purpose and storytelling can play a key role, and which we explore in much more detail below. The final criterion is about actually having the enabling technology – which I couldn't agree with more, but which is not within this book's sphere of expertise – but then also, having acquired the right technology, using it as an integral part of doing business. Here, too, purpose and stories have a role to play and so we also touch on this below.

Building Networks

Networks – business ones and social ones – can, and usually do, happen naturally. We are social creatures who naturally seek out community. But an organisation, or indeed an individual, that is committed to building agility needs to learn how to proactively design, build and nurture networks of different types, and then how to use them effectively.

The reason that networks are identified as a key factor in agility is (presumably, though no one seems to be spelling this out) because they strike that balance between autonomy and accountability *and* enable ideas, knowledge, skills and resources to be accessed, shared and deployed in a way that solves problems or responds to opportunities faster and better than a more traditional model does.

There is clear, bottom-line value to be found in getting this right. For example, in her work on collaboration, Heidi Gardner presents a compelling business case for larger networks, well used, by showing a clear correlation between a larger network and higher revenues amongst partners in a professional services firm.[4]

But there is also the potential for a terrible amount of value and time to be wasted if we try to build networks and collaborate with one another just for collaboration's sake. This starts to look like Having Lots Of Meetings and Answering Lots Of Emails and never getting any Actual. Work. Done. It also risks actively cannibalising the knowledge, skills and discretionary effort that would otherwise exist within an organisation because, as shown in a powerful piece of research by Rob Cross and colleagues,[5] the same very few people in an organisation get sought out time after time for their knowledge, support and time, and over time this group becomes less effective and less engaged.

So, what does good look like? Swoop Analytics, an organisation which analyses social networks, has developed a model based on their work across a wide variety of industries and geographies, which maps network performance based on cohesion, on one axis, and diversity, on the other, and concludes that the highest performing networks are both highly cohesive and highly diverse. These findings dovetail perfectly with Sam Coniff Allende's observations[6] that the crews assembled by the Golden Age pirates were profoundly successful (albeit in pursuit of slightly dubious purposes) because they were diverse, inclusive teams, strongly bonded and yet able to scale up and scale down and move from command-and-control to collective decision making and back again as the circumstances demanded. Let's consider then what practical steps can be taken to move a 21st-century organisation up-and-right in the ubiquitous 2x2 matrix. How can they be more pirate?

Cohesion

Thinking about cohesion first, then, part of this is simply about people getting to know one another and building trust. There is lots of theory on this in Chapter 9 on Belonging, and some examples of how to do this in practice, drawn from CMS and elsewhere. In particular, it is worth recalling Seth Godin's work on tribes, which talks about the need for a narrative, a way to communicate and something to do.[7]

There are also some more overt structural and policy changes that organisations can consider making in order to encourage the development of effective networks, including implementing clear, flat structures which cluster people into 'performance groups' – being mindful of Dunbar's Number, which says that the largest effective group size amongst human beings is around 150 people[8] – having clear, accountable roles, and a hands-on approach to governance,[9] and ensuring that the physical environment encourages the building of relationships.

In some respects, law firms tend to perform quite well on this cohesion dimension of networking. Swoop Analytics has found that partner-led professional organisations tend to fall squarely into the 'tribal' box – bottom right – because they are highly cohesive, but low on diversity.[10] At CMS, its recent merger of course meant that the team had to work harder on this aspect than it might otherwise have in a theoretical 'steady state'. We have talked at length about the belonging piece, but framing this in network terms, the leadership team's focus had to be on building cohesion across the legacy firm boundaries – breaking into, without breaking down, the old networks that saw old friends attending the firm's carol service together, or booking flights for the conference together, or bantering together on email late at night.

CMS, also gave serious thought post-merger to some of the more structural aspects around building cohesion. Partly, I think, this was because the merger required that some of the things which commonly go unspoken in law firms – basic things like, 'Whose team am I in?', 'What is my job?' and 'Who gets to decide?' – be made explicit.

Again, in a steady state firm, it is not that people don't know the answer to these questions, it's just that lawyers' high need for autonomy, coupled with a degree of ambiguity or nuance around some of the answers, can make it easier not to go there. In the context of change, it becomes even more important to be explicit. It is hard for people to figure things out at scale in an unfamiliar context and so it becomes important to help people by putting clear structures and roles in place

Diversity

The other dimension on Scoop Analytics' 2x2 is diversity. We are talking here about diversity in the broadest sense – including how far a network reaches out beyond its own echo chamber – and the thinking is that a network which can effectively bring lots of different ideas and experiences and perspectives to bear to solve problems or take opportunities will, ipso facto, be a diverse network.

In the case of the CMS merger, all three firms had a deep and long-standing commitment to maintaining an inclusive culture and driving to increase diversity in terms of gender, race and societal background – indeed this was one of the areas in respect of which of the three firms' cultures were most strikingly aligned. Nevertheless, law firms are populated by lawyers, and some other business professionals, who tend to be referred to as, well, 'non-lawyers'. Ditto engineers, architects, consultants, though less explicitly so, and accountants perhaps less so. The point really is simply that in a professional services context, there is a certain homogeneity of thought and skill set, that can mean that such organisations – even those most deeply committed to inclusion – can sometimes struggle to bring to bear in their networks the diversity that they keenly need, given the complexity of the work that they do. At CMS, the merger conferred one additional huge advantage in this regard in that the new firm enjoyed the further diversity brought about by bringing three cultures and histories together. The new firm also, of course, put diversity and inclusion at the heart of its strategy and ensures that it measures and rewards against 'we include' as a key behaviour right across the business.

It's Called netWORK for a Reason . . .

It can be useful to identify in your own organisation where on that 2x2 matrix you are starting from. At CMS, the leadership team gave Heidi Gardner's *Smart Collaboration* as required reading to the whole partnership to give them some sense of what 'good' network-building and collaboration could look like in their context, and to make the case for what it could deliver. Building on her body of research on professional services firms CMS then commissioned Gardner and her team to conduct some research on CMS specifically, looking at the extent of collaboration, the nature of the untapped opportunity and any perceived barriers to collaboration. CMS then worked with the team to design a series of skills-building workshops for leaders to help them inculcate a culture of network-building and collaboration across their teams.

At the time of writing, there is some interesting different research under way in other professional services organisations which aims to map some of the key networks which exist within the firm and see what can be learned about how they work, and what value they deliver. The ultimate vision is of a much more porous organisation, where divisions between specialists, and between the lawyers and the nons, are eroded, and people come together in fluid communities of knowledge, skills and practice in order to meet client needs and explore new opportunities, with real ease, clarity and efficiency.

Where do Purpose and Stories Fit In?

How does having a clear purpose and a strong narrative help with the development of networks? Well, we have already discussed at length the role of purpose in building belonging. A clear sense of purpose can also help with 'legitimising' some of the stickier systems and process aspects of cohesion discussed above. In my experience, it is easier to nudge an organisational culture towards being more accepting of a more overt approach to setting roles and boundaries once the link to our purpose is made explicit.

A common purpose helps to build cohesion within a diverse team, and stories help to build both cohesion *and* diversity, by building understanding of different perspectives and helping us to see common ground and shared experiences. This in turn builds trust and respect.

In practical terms, I have found that one of the best ways to build relationships and confidence and to boost morale is through giving people the opportunity to collaborate with one another; and a close second is by sharing real life stories of real life collaboration generously, extensively and repeatedly across the organisation. CMS used collaboration stories at its landmark conference, six months post-merger, to showcase examples of best practice and share a vision of what a well-networked, collaborative CMS could look like.

Getting to Go

The third aspect of agile organisations that McKinsey identifies is that of rapid decision making and rapid learning. This is about both how organisations think, and how they act, and getting it right requires organisations to be focused on performance rather than process, to standardise the process stuff (meetings, formats of documents, briefing protocols) so that it helps rather than hinders, and to ensure complete transparency of information. Above all, though, it requires organisations to be able to generate ideas, create prototypes, experiment and learn from their experiments – both their successful ones, and their failures.

Organisations that get this right have an advantage because they manage to bridge what Ali Tisdall, of the leadership and applied neuroscience consultancy Mind3, calls, 'the valley between strategic intent and operational implementation'.

They know how to get things done. Frederic Laloux[11] says that instead of a strategy document, purpose-driven organisations have:

> . . . a very clear, keen sense of the organisation's purpose and a broad sense of the direction the organisation may be called to go.

Laloux argues that a formal strategy would limit the possibilities to a narrow course – a bad idea in a VUCA world – whereas, if an organisation is truly purpose-driven (and purpose-designed, I would add), strategy instead happens organically, all the time, all over the business, as people test things out in pursuit of the purpose, confident of the framework within which they are working. Good ideas then catch on and gain momentum, and others fall away. This has echoes of the coalface argument in Chapter 10 on Evolution – people feel empowered to act locally, and are intrinsically motivated to do so because they have ownership of that part of the story; they are leaders.

This all sounds wonderful, and makes perfect sense – provided that in the absence of a strategy, organisations do indeed feel empowered to think and act, and don't instead just sit frozen, like rabbits in the headlights and wait for someone else to make the first move. Often, the biggest obstacle for organisations to overcome in order to get into a position where they really are living in a rapid decision-cycle is a psychological and behavioural one around creating without self-editing, and experimenting and taking risks.

This is perhaps a bigger challenge for lawyers than for most. Lawyers are trained to identify, minimise and manage risk. They like to kick the tyres on proposals, consider every possible scenario, and absorb every piece of possibly pertinent data before taking a decision It takes a profound shift in mindset to ask them just to give something a go. 'What's the worst that can happen?' you may say. A lawyer is just the person to tell you! But this difficulty extends well beyond lawyers. We were all of us raised in an industrial age or in its death throes, and educated in a system that taught us that there is a right way and a wrong way to do things, and that if we focus on process and follow a series of steps we won't go far wrong. It is a stretch for any of us, in that context, to take the creative approach that McKinsey advocates.[12] If we want to make something of Leroux's vision then, we need to dig a little deeper into what is stopping us.

Something Related to Fear

In *Daring Greatly*,[13] Brené Brown recounts a conversation with Kevin Surace, the then CEO of Serious Materials, in which she asked him:

What's the most significant barrier to creativity and innovation?

Surace identifies 'something related to fear [that] keeps people from going for it'. He says:

The problem is that innovative ideas often sound crazy and failure and learning are part of revolution. Evolution and incremental change is important and we need it, but we're desperate for real revolution and that requires a different type of courage and creativity.

Perhaps we are afraid. Afraid to take risks, perhaps afraid to deliver poor service, or to cost our organisation money, and above all afraid of being exposed, looking stupid and damaging our reputation and standing.

If that's the nub of the matter, what can we do? We can do some work. Individually, we can learn to recognise and master the fears that grip us.[14] And as leaders in our organisations? Brown concludes that:

to reignite creativity, innovation and learning, leaders must re-humanize education and work.

Brown says that in order to re-humanise, engaging with vulnerability is vital. She cites an interview with Gay Gaddis, the owner of T3 – the US's largest advertising agency owned by a woman. Gay says:

By definition, entrepreneurship is vulnerable. It's all about the ability to handle and manage uncertainty. People are constantly changing . . . you have to stay nimble and innovative. You have to create a vision and live up to that vision. There is no vision without vulnerability.

So, vulnerability again . . . In my experience, one of the best ways to engage with vulnerability is to ask questions. Say 'I don't know how to . . .'. Actively enlist help and seek ideas. If we can do this as leaders, others will follow. Try things. And tell stories. Tell stories about the craziness and failures that are part of innovation. Showcase courage and creativity, and make the journeys and struggles themselves part of an organisation's story. This will enable the power and potential of purpose to catch fire, because people won't be waiting to act. People will be able to overcome their fear, create and share ideas, and harvest the learnings from successes and failures alike. Put systems in place that can churn through lots and lots of ideas quickly in order to find the good ones, and rig the whole thing to remind people that pain and frustration and bad ideas are not some kind of aberration, but rather a vital part of the process of creativity. Focus on the people and the process and on nurturing a strong and effective team dynamic, not the ideas or outcomes themselves.

The People Model

The fourth aspect of agile organisations that McKinsey identifies is a dynamic people model which engages and empowers people, fosters a strong community, and builds the skills for innovation and agility. We have talked a bit in Chapter 9 on Belonging about how to foster community in organisations, and will dig deeper into some aspects of that in Chapter 13 on Understanding, so let's focus for now, first on the piece around engaging and empowering people and, secondly, on building the skills for innovation and agility.

Engage and Empower

If the aim is to increase engagement and empowerment, there is ample research evidence to suggest that a purpose-led approach is a good one. Daniel Pink, in *Drive*,[15] sees the purpose motive express itself in organisations in three distinct ways – first, in the use of profit to achieve purpose; secondly in a corporate rhetoric that emphasises more than self-interest; and third:

in policies that allow people to pursue purpose on their own terms.

Birkinshaw and Ridderstråle argue in *Fast/Forward*[16] that the most agile firms are those that can link decisive action with emotional conviction. This is reminiscent of the Japanese proverb which says:

Vision without action is a daydream; action without vision is a nightmare.

Purpose is what underpins both vision and emotional conviction. It both gives the intrinsic motivation to act, and guides the action. Birkinshaw and Ridderstråle point out how fickle we human beings are – how prone to forget about, or wander from, purpose-goals in pursuit of pleasure or our own gain. Purpose therefore needs supporting systems – formal or informal, but consistently applied – in order to stick. The people model of an organisation would be one such supporting system. Another would be that which develops when we incorporate tangible examples of an organisation's purpose into everyday life for everyone in the business. Storytelling is a key – perhaps the primary – means by which this can be achieved.

Permit me a short hop from the storytelling metaphor into the visual arts. Mark Humphries describes a workshop in which he and around twenty other senior leaders of an organisation he was working with sat in a circle. A glass jug of orange juice was placed in the centre of the circle, on a rumpled blue cloth. Each member of the group was given a piece of paper and some crayons and asked to draw the still life in front of them. A few minutes later, the group shared their pictures: twenty completely different pictures of the same tableau. Each was drawn from the artist's physical position in the room, and so was different in terms of actual physical perspective, but each was also completely different in terms of interpretation and expression.

This is a metaphor (we're knee deep in them here) for how purpose and storytelling can work in an organisation. You place your purpose at the centre – clear, concrete, an actual thing – and you give people some tools to help them – examples, role models, stories . . . and then you Leave Them Alone to engage with the subject matter and make sense of it for themselves. You give them the agency to interpret things in their own way, and to tell their own story, within the guiding context of the purpose.

Building Skills

The point around building skills for innovation and agility is exciting. Yes, some of the change required goes quite deep into behaviour and mindset and needs work, but there are also some skills that can be readily taught and learned that will make a big difference. It is possible to teach leaders how to collaborate and to help them to foster even more vibrant networks. It is possible to teach people design-thinking to help them to generate ideas, assess them and develop prototypes. It is possible to leverage tech to give people a place to share ideas and build on and stress test the ideas of others. Teach people to code, and give everyone a working knowledge of how technology can make a difference. Showcase examples of approaches to innovation from across different sectors – learning from start-ups about how they develop and test ideas, and holding hackathons and 'Dragon's Den' type events for clients in different sectors. Encourage leaders to consider taking on non-exec or advisory roles[17] to broaden the organisation's horizon and perspective by bringing stories, and experiences from the outside into the organisation. All of these things help people to build the skills and perspectives that can help them to innovate.

A Fool With A Tool . . .

I'm a sucker for a pithy little aphorism, and for anything in rhyme, and so was delighted recently when a friend who is a senior in-house lawyer, when talking about his frustration with law firms and their tendency to talk about AI loosely and liberally as if it's the answer to everything, summarised his view by quoting the American software engineer Grady Booch:

A fool with a tool is still a fool . . .

The fifth aspect of agile organisations identified by McKinsey is that they embrace enabling technology as an integral part of how they do business. The key words in this sentence are enabling and integral. Technology *per se* is unlikely to be a source of competitive advantage for any organisation except a technology company. Apart from anything else, most of us are buying the same tools from the same developers. Application is everything. Law firms don't sell, and clients don't buy, information about what the law is. Law firms don't sell, and clients don't buy, clever pieces of tech that make it quicker to figure out what the problem is. Rather, firms sell, and clients buy, an understanding of how to apply and deploy information and technology, and the ability to pull together numerous complex strands in order to design a solution or create an opportunity. Technology can help with this, and agile organisations link this back to their purpose.

Bringing It Together

To conclude, then, because this chapter has had a few different strands:

- There are some paradigm shifts in societal and business models and in the availability of information and technology which are driving organisations to become much more agile.

- When organisations are changing, either directly in response to the above pressures, or because there is another catalyst for change, they should be looking to design in as many aspects of agility as possible.

- Agility balances stability and dynamism – like a gymnast's strong spine.

- McKinsey identifies five defining features of the most agile organisations: they are purpose-led; they have dense networks; they have rapid decision and learning cycles and experiment; they have people models which support agility; and they have enabling technology as an integral part of how they do business.

- We can and should take practical steps to buildin all of these. At least in the legal sector, and possibly for the majority of us, one of the hardest aspects of the features listed by McKinsey to develop is the willingness to experiment and take risks. This requires us, as individuals and leaders, to address issues around fear and vulnerability.

• Purpose and storytelling serve as both the strong spine and the enabler of dynamism in agile organisations.

Notes

1. Whyte, David, 'Working Together' from *River Flow – New and Selected Poems*, Many Rivers Press, 2012.
2. Aghina, Wouter, De Smet, Aaron, Lackey, Gerald, Lurie, Michael and Murarka, Monica, 'The Five Trademarks of Agile Organizations', McKinsey, January 2018.
3. In *A Good Time To Be A Girl*, Helena Morrisey points out that such an approach can 'tend to favour women's ways of working and behaving' – further evidence, as if more were needed, to support the argument that a diverse and inclusive approach at all levels in an organisation is imperative for agility. Morrissey, Helena, *A Good Time To Be A Girl*, William Collins, 2018.
4. Gardner, Heidi, *Smart Collaboration: How Professionals And Their Firms Success By Breaking Down Siloes*, Harvard Business Review Press, 2017.
5. Cross, R., Rebele, R., Grant, A., Collaborative Overload, *Harvard Business Review*, Jan–Feb 2016.
6. Coniff Allende, Sam, *Be More Pirate*, Penguin, 2018.
7. Godin, Seth. *Tribes – We Need you to Lead Us*, Piatkus, 2008.
8. Dunbar, Robin, *Human Evolution*, Pelican, 2014.
9. Aghina, Wouter, De Smet, Aaron, Lackey, Gerald, Lurie, Michael and Murarka, Monica, 'The Five Trademarks of Agile Organizations', McKinsey, January 2018.
10. Lock Lee, Laurence, Smart Collaboration = Smart Money, www.swoopanalytics.com blog, January 2017.
11. Laloux, Frederic, *Reinventing Organizations: A Guide to Creating Organizations Inspired by the Next Stage of Human Consciousness*, Nelson Parker, 2014.
12. I find this a real frustration with so much of the business literature out there – it might be exquisitely researched, distilled and presented, but it is of limited practical relevance if no one is prepared to get down and dirty enough to really tackle the 'how'.
13. Brown, Brené. *Daring Greatly – How the Courage to be Vulnerable Transforms the Way We Live, Love, Parent and Lead*, Penguin, 2012.

14. For help, see for example Brené Brown's *Daring Greatly*, above, *Taming Tigers* by Jim Lawless, and *Playing Big* by Tara Mohr. All very different; all brilliant.
15. Pink, Daniel. *Drive: The Surprising Truth About What Motivates Us*, Canongate, 2009.
16. Birkinshaw, Julian and Ridderstråle, Jonas, *Fast/Forward: Make Your Company Fit For The Future*, Stanford Business Books, 2017.
17. Subject of course to their client commitments, which remain top priority, and the need to ensure no actual or perceived risk of conflicts.

Chapter 13

Understanding

Each of us is here for a brief sojourn; for what purpose
he knows not, though he sometimes thinks he senses it.
But without deeper reflection one knows from daily life
that one exists for other people.

Albert Einstein[1]

This part of the book is brought to you from Hjørring, one of Denmark's oldest towns, on the most northerly tip of Jutland, with a population of around 25,000 people. I am here for a week, and for a number of reasons it seems a good place to be writing about communication and – more broadly – about how we understand one another.

The people here, unsurprisingly, speak Danish. I don't. I have just been to a truly beautiful supermarket, and navigated my way through towering aisles full of food with names and labels that I am unable to read. Of course, because most of the products are familiar, I have come back to my apartment with everything I need. But because I am insatiably curious, I have been looking up lots of the words that I saw. So, for example, I looked up 'Sild', which I saw written all over the place. Herring, literally. More generally, fish. But also used colloquially here the way, in the UK, we might use another metaphor. Squashed in like sardines on the tube? You'd be squashed in like herring in Hjørring (though rush hour feels a little less arduous here). Been told your idea is worth peanuts? Come to Denmark, it's worth a whole herring. Worried your career is a dead duck? It's a dead herring here. I love this, and how the metaphors people use can give us such rich insights into the lifestyle, history and culture of their place in the world.

I've mentioned I don't speak Danish. But almost everyone I have met here speaks exquisite English, or at the very least, will have a good go at it. This makes me feel, above all, very grateful, as it means I can navigate my way through the various things I need to do every day. But it also makes me feel vaguely guilty, that it is other people who are making all the effort and not me, and vaguely ashamed and disempowered that I don't have the knowledge and mastery of a particular skill that I would ideally like to have in order to more fully participate in life here.

I am here because my son is participating in the DanaCup – a youth football tournament involving 1,100 teams of young people aged between 10 and 18 from forty-five different countries. When opposing teams meet, they rarely speak the same language as each other, or – perhaps more significantly – as the referee. This means that a lot of communication – both formal and informal – is via sign language and gestures.

My son is the goalkeeper for his team. Now, the uninitiated amongst you might assume, as I did for a long time, that the extent of his role is to stand around for most of the match and then – hopefully – stop any shots on goal that come his way. But there's much more to it than that. It is my son's job to 'run the defence' – that is to give direct commands to his team mates in defence about their position on the field and their tactics. Increasingly, as he gets older and more experienced, he is also expected to play a role in directing the overall on-field game strategy, because he can see the entire pitch and how the game is unfolding more clearly than the other players can. This means that my son has to be really decisive and really assertive. He has to work out in the moment what he needs a player to do and then he has to tell them – sometimes one to one, and sometimes yelling the length of the pitch, over the cacophony of parents cheering their encouragement and hollering their own opinions.

There are a lot of lessons to be gleaned about communication and understanding in all that!

What Do We Mean by Understanding?

In this chapter, we will explore the meaning and importance of understanding, and think about the various different levels at which we communicate and connect with one another – from barked instructions in the heat of the moment through to a much deeper meeting of minds that transcends culture and language.

Patrick Lencioni in *The Advantage*[2] presents a simple four-part model for organisational health, and at the heart of his model is the need for clarity. I agree that clarity is critical and it is going to be one of the key concepts that we explore in this chapter as we think about how we communicate, but as a single overall concept and title for this chapter, I prefer 'understanding'. Understanding has both cognitive and emotional dimensions. There's a nuance in the concept of understanding that is about relationship. It carries with it the notion of acceptance, a sense of generosity and openness. It is about a deep knowing of who and what we are as an organisation, and between one another as individuals about who we are, how we roll and what we contribute.

'Send Three and Four-pence We're Going to A Dance'[3]

In order to understand why understanding matters consider the mounting pile of carnage which develops when communication doesn't work well and understanding isn't present:

- there's the frustration in the moment – for either or both of the communicator ('Get to the halfway line, Josh!') and the communicated-to ('If that's what you wanted me to do, why didn't you tell me!');

- there's inefficiency – I might have to tell you something twice, or do something myself that you could have done better;

- it can lead to wasted effort, duplication, dropped balls, missed opportunities ('I thought YOU were calling that important potential client . . .');

- it can create unnecessary risks and lead to a drop in quality and to mistakes;

- you miss out on the unknown unknowns – the good idea that never was because the two people who each held half of the idea could not communicate with each other;

- poor communication can actively drive a wedge between people ('She never listens!' vs 'He never explains what he wants!');

- people can get overwhelmed by the sheer volume of information coming their way;

- this leads to a risk that they will miss something important;

- it also means that people don't feel heard, understood or valued;

- over time this leads to people being disengaged . . .

- . . . and a lack of cohesion in teams

- . . . and all of the above leads to untold destruction of value in an organisation.

During periods of change, all of the above can be particularly acute for a number of reasons. People may lack familiarity with each other, and so we can't use all those shortcuts and awareness of one another's proclivities in order to add clarity to an unclear message. Levels of trust may be low and people may perceive a lack of transparency and fear that information is being kept from them. On the other hand, the sheer volume of information to be communicated is likely to be much greater than usual. Finally, an organisation may already be on the back foot in terms of rolling out unfamiliar systems and processes, or in terms of building engagement and cohesion, and so the impact of a wrong note in terms of communication has a much greater adverse effect.

In every change project I have been involved in, communication has been, if not quite 99% of the challenge, then a very significant issue. Communication is really hard to get right, not least because different stakeholders tend to want very different things. That said, the good news is that there are several practical steps that can be taken to ensure good communication and help to build understanding. Some of the practical steps that worked well for the team during the CMS merger, I have already outlined in Part Two – the aim in that case was to create and implement a robust and intelligent communication strategy to build understanding and cohesion and to oil the wheels of the changes that were being made.

In the rest of this chapter, let's dig a little deeper and in particular explore how a sense of purpose and the ability to tell stories are absolutely fundamental to the development of understanding in an organisation. The prize for organisations that get to grips with understanding is high – greater ease of execution, faster decision making, higher levels of engagement and participation, and the ability to tap into collective intelligence.

Two Ears, One Mouth[4]

But first, it is important to remember that there are two people in any communication scenario: the speaker and the listener; the writer and the reader; the communicator and the communicated-to. Before we consider how the communicator builds understanding by conveying information and ideas effectively, let's consider the equally active and important role of the recipient of a piece of communication. What does it mean, as an individual, a team or a whole organisation to be a good listener?

Any parent or anyone who has been around small children will be familiar with the particular frustration that descends when your beloved little one, rather than responding to a simple instruction – 'Put your shoes on, Frankie' – instead sticks her fingers in her ears and yells something like 'Haha! I'm not listening! I can't even HEAR you!' Crude, but effective, and in truth I suspect that lots of us are like this a lot of the time.

There are all sorts of things that get in the way of our ability to listen well (I'm using 'listen' here because I believe verbal communication to be paramount, but it is shorthand for 'receiving information and ideas' and includes, for these purposes, reading too). Often, we are simply completely overwhelmed by information – there is too much, its relative importance, urgency and relevance is not clear, and there is no context. It's just noise. Sometimes, our own internal muddle or overwhelm means we can't receive any more incoming information. Sometimes, we bring so much of our own stuff to the table – our prejudices, biases, sense of hier-archy, fixed mindset, pride and desire to be right, disengagement, defen-siveness, fear – that we cannot hear through it. Sometimes we are not even trying to listen – we're just waiting to speak. Sometimes, we are in conflict and are too emotional to be able to listen – we know that there is a strong emotional element in listening, because we know that we listen much bet-ter when we are in love with someone.

Happily, even when we're not in love,[5] listening is a skill that can be learned and practised and improved. Simply increasing our amount of eye contact, for example, will provide an immediate uptick in the quality of our listening and in the amount of information we retain and are able to recall later when tested.[6] There are also a number of traits of empathetic listening which can be practised, including:

- putting the feelings and thoughts of others first;

- letting your guard down and being open about your own emotions and opinions . . .;

- . . . while also being able to imagine yourself in the experience and perspective of others; and

- avoiding judgment or criticism.[7]

This is all undeniably really difficult. Particularly when we are busy, moving quickly, and heavily invested in what we believe to be the right way of doing things. It is particularly hard when what we are being asked to listen to is feedback on us, and our own performance. Hearing the phrase, 'Can I offer you some feedback?' sets off exactly the same stress reaction in our brains as being chased down a dark street at night![8] We perceive a threat to our psychological safety and we react accordingly. We need to be mindful of this and learn to manage our response, stay curious about the information we are receiving, and remain open and receptive.

Organisationally, we need to think about what structures and processes we might need in order to help us to listen well, as well as ensuring that our culture remains open and receptive. CMS, for example, had various email addresses that people could send feedback to, and a physical 'suggestions' box in the restaurant for those who preferred to remain anonymous. The firm uses crowd-sourcing software to enable people to share ideas, and encourages the use of team discussion platforms. The leadership team conducted a number of set-piece listening sessions in the run-up to the merger and since, and there are a number of ad hoc groups that the leadership team can use as sounding boards. This remains a work in progress – the firm is committed to ensuring that everyone has a voice.

There is much for us to learn then, about how to be better listeners, as individuals, as leaders, and collectively as organisations. Mastering this, or even getting just a little better at it, will deliver value both because being listened to *per se* makes people feel engaged and valued, and because listening well puts you in receipt of invaluable information, perspectives and insight which you can then put to use; be it to reconfirm the path you are on, cause you to alter it, or cause you to adopt a new approach to building buy-in.

But as we come now to consider the other side of the coin – i.e. how to be effective speakers and writers – it is important for us also to bear in mind that the vast majority of people with whom we are communicating are not very good at listening either. They are just as prone to the same roadblocks as we are, and good communicators have to recognise this and adapt accordingly. This means that we need to operate in a realm beyond logic and efficiency – we are not working in a widget factory. As Sir Ken Robinson in his book, *Out of our Minds: Learning to be Creative*, writes:

> However seductive the machine metaphor may be . . . human organisations are not actually mechanisms and people are not components

in them. People have values and feelings, perceptions, opinions, motivations and biographies, whereas cogs and sprockets do not.

How to Communicate

You're Doing It Anyway

The first point to note when thinking about communication is that we are doing it anyway. Already. Right now. It's just that sometimes, we use words. The word 'communication' literally means 'to share' or 'to make common'. All of us, individually (unless we are hermits) and as organisations are sharing something of ourselves all the time. The decision not to actively communicate something, or to prioritise communicating one thing over another is, in itself, an act of communication.

There's an old marketing aphorism which says that 'You are the message', but I prefer Rob Bell's take: 'You are the medium',[9] through which the message flows. This means we need to think hard about integrity and consonance: we need to practice what we preach; and if you're going to talk about detailed points of technology or process one day and big picture strategy the next, there had better be a clear thread that holds all that together. Purpose-led organisations (and purpose-led people) find this much easier to pull off than others because they already have that central guiding idea as an integral part of how they do business, and are intrinsically motivated to act and speak in accordance with it. It also becomes easier to communicate across a range of subject matters without creating confusion because there is a unifying narrative thread.

Do You Have To?

Okay, so assuming you think that you have got something that you actively want to communicate with words, written or spoken, the next thing to do is to test whether communicating anything at all is a good idea. Please don't misunderstand me – as a general proposition I am firmly on the side of transparency – but given that we are generally drowning in information, being very deliberate about what information needs to be communicated, when, how and to whom is a critical first step on the path to communicating effectively and thus building understanding.

Perhaps the best test for whether one should communicate is the high bar set by the prophet Jeremiah when he explains why, despite terrible setbacks and opposition, he is going to continue to speak about God:

> If I say, 'I will not mention him, or speak any more in his name,' there is in my heart as it were a burning fire shut up in my bones, and I am weary with holding it in, and I cannot.
>
> **Jeremiah 20:9**

Do you have a burning fire shut up in your bones? Really? Applying a test like that will ensure passion and – back to the 'you are the medium' point – for all but the most prosaic types of communication (see below), the impact will be much more powerful if the message is delivered with authentic feeling.

If your piece of planned communication fails this test, perhaps now is simply not the right time? Or perhaps it needs to be packaged up in a different way with other things to make something more compelling? Or perhaps you need to think differently about *how* it should be communicated – is it something for the intranet? A future presentation? An email?

Do YOU Have To?

Perhaps there is a message that certainly needs to be delivered, and yet for whatever reason you are not the person who most feels the passion in relation to this message. In this case, it is worth considering whether there is someone else better placed to deliver the message.

Another aspect to think about when considering who is best placed to communicate a particular message is personal brand, role prominence and/or seniority. Even in flat structures, the most senior voices can carry significant authority and influence. People like the most significant news about their organisations and their careers to come from the most senior person.[10]

A final thing to think about in terms of who should communicate is how important it is that a message is heard directly from you, or whether it can be cascaded down through the organisation, with people hearing locally interpreted versions of the same message. There are advantages to both, and a cascade approach is often more efficient and allows space for tweaking the message to suit the audience. But it comes at a cost, as something is always lost in translation around nuance, tone or even accuracy. For a lovely visual illustration of this last point, have a look at a very short film by Clement Valla which illustrates what happens when a line is traced 500 times . . .[11]

What Are You Trying to Do?

Assuming you have got something to say, and have established that yes, it needs to be said now, and by you, the next thing to think about is the aim of your communication.

In my mind's eye, I see almost a spectrum running from 'clarity' at the one end to 'enlightenment' at the other. Rob Bell talks about the difference between transmitting (straightforward information, coming straight from my head to yours), translating (pulling various strands together and sense-making – presenting an argument, or seeking to persuade), and transforming (where my aim is a wholesale shift in your outlook, most likely also involving a degree of emotional engagement).[12] These different types of message, these different aims, would indicate a different approach. Is your aim to ask or encourage or enable people to actively do something?

Or are you trying to work at a deeper level, to perhaps shift attitudes, and create a different culture? Again, a different approach will apply.

For me, the more deeply interesting part of this, for the purposes of a chapter about engendering understanding, is the piece towards the right of my spectrum; thinking about communication as a tool for enlightenment, transformation and the development of culture. But the piece around clarity is also enormous and can be hugely challenging during a period of change, so let's tackle that first.

Clarity

Why does clarity matter? At the most basic level, during periods of change, there is just an enormous amount of information which needs to be conveyed. This is just plain old hard work for everyone's brains, and so the clearer the information can be the better. But clarity also helps at a deeper level by reducing uncertainty and so helping the brain to avoid being in a threat state. At times of change, there is lots of uncertainty and ambiguity and we find this stressful. By scotching rumours, 'showing your working' so that people understand the process, and reducing uncertainty even a little – 'We don't know yet, but we will know on Tuesday when we will know' – clear communications can reduce stress and so increase focus and engagement.

In the second week of our new firm's life, in response to a handful of queries about how the overall structure of CMS worked, we offered to hold a 'lunch and learn' session in one of our seminar rooms in London. We ordered some sandwiches and laid out some chairs, as you do, and we had a few slides because structures and reporting lines are easier to explain with pictures than with lots of arm waving. Three hundred people turned up on less than 24 hours' notice. It felt a little like running a church fete. We opened up the partitions to the adjoining rooms, ordered more sandwiches, got more plates. Someone stood on a chair and arranged people into orderly queues.

I was worried that we had somehow mis-sold the session. Were they expecting a big announcement? A celebrity? No. There were just lots and lots of people, new to the firm, and with an enormous appetite for information and understanding. They wanted to know how things worked and how they fitted in. Clarity matters.

Lencioni[13] – who, you will recall, says that clarity is critical for organisational health – goes on to say that that there are six critical questions which leaders must be able to answer and that all leaders in an organisation must be able to answer them in exactly the same terms:

there can be no daylight between leaders around these fundamental issues.

The first question is: 'Why do we exist?' That is, of course, *the* purpose question.[14] From there. Lencioni says that the next question is, 'How do we behave?' and the third question is, 'What do we do?'

These three questions, in this order, chime with Sinek's Golden Circle, which we considered earlier – layering values and culture, and then strategy, onto purpose. Thereafter, Lencioni's questions are more tactical and structural: 'How will we succeed?', 'What is most important, right now?' and 'Who must do what?'

In a nutshell, Lencioni says that we need clarity on purpose, values and culture, strategy, tactics, priorities and roles. I could not agree more. Purpose is front and centre again in its own right, but it also permeates the others and helps to bring the clarity that Lencioni seeks:

- we have already discussed the relationship between culture and values and purpose;

- purpose is also the underlying piece which will enable organisations both to be evolutionary and agile in their setting of strategy and tactics and then to 'make sense' of strategy and tactics as they inevitably change in response to market conditions;

- we have talked, too, about clarity on roles, and how a clear sense of purpose incentivises and legitimises conversations around who does what which may otherwise be difficult in a consensual culture; and

- purpose also ensures a deep understanding of how any given role 'fits in' and helps with intrinsic motivation and engagement.

Storytelling also has a role to play in bringing clarity – in particular, in helping to manage expectations and normalise experiences, and in minimising uncertainty and ambiguity in inherently uncertain circumstances. If you can look at how other people have navigated a similar experience, what they have learned and how things unfolded over time, that can shed a light on your current experience and bring a degree of perspective and insight that it can otherwise be hard to grasp hold of in the heat of a particular moment.

Enlightenment

We have some sense, then, of the various factors which can help to bring the clarity part of understanding to a changing situation, and how purpose and stories can help us to achieve that clarity. But what does it take to move beyond clarity and to help people to make deeper sense of their situation and to understand one another – perhaps even to effect a shift in attitudes and perspectives? How do we enlighten one another about the mysteries of who we are, and discover what we could be together in an organisation? And what might purpose and stories have to offer us in this context?

Again, the starting point for effective communication is to remember that we must meet our audience where they are. None of us is a blank page – we interpret things according to how we understand the world to be based on our life experiences so far. This means that people will hear what you say through the prism of what they expect to hear, and this in turn is based on the particular experiences, biases, fears and assumptions they are carrying into the room with them.[15] It is perhaps worth remembering that, while this might make communicating more difficult, we actively and absolutely want this to be the case. We want a diversity of backgrounds, experiences and perspectives on our team, and we want people to be able to be fully themselves at work.

Amongst this group of people, then, each bringing their own 'stuff' into the room, the first requirement is to build trust – vis-à-vis each other, and vis-à-vis the organisation. This is far from just a fuzzy nice-to-have; we know that investing in the connectedness between people in a team both increases their productivity and reduces risk. Groups who are socially sensitive – who are tuned into one another and each other's needs and can pick up on subtle changes in mood – are higher performing than groups who are less socially sensitive.[16]

Erin Meyer[17] makes a distinction between two different types of trust. Cognitive trust – trust from the head – is based on another person's accomplishments, skills and reliability. This is relatively easy to establish between people, at least in a professional services context, where a person's quality and standing is generally assumed, or at least their reputation is theirs to lose. It is also relatively easy for an organisation to support and encourage the development of this sort of trust – by sharing information about people's credentials and achievements, such as the daily Meet The Partners email we distributed at CMS in the early weeks of the merger, and by giving people an opportunity to work together and test each other's skills and reliability in 'real life'.

The second type of trust that Meyer identifies is affective trust or what she calls 'trust from the heart', which is based on feelings of emotional closeness, empathy and friendship and – of course, of course – depends on people being able to display a degree of vulnerability with one another.

Again, as we saw in Chapter 9 on Belonging, this type of connectedness can be built and is not, in truth, terribly difficult to achieve but it does require time and commitment and an openness to learning.

It is so easy to misunderstand each other! For example, Meyer talks about peach cultures and coconut cultures – a somewhat crude but useful distinction between national cultures in which people are ostensibly very open and warm on first meeting but may be hard to get to know more deeply, and those in which people are cooler at first, but establish deep relationships once trust has been established. Country of origin notwithstanding, there are peaches and coconuts in every team, and on first meeting coconuts can perceive peaches as superficial and hypocritical, and peaches can consider coconuts to be stand-offish and rude.

Building Trust and Getting Enlightenment

Getting past these initial misunderstandings in order to build trust and enable a team to communicate effectively at this deeper level takes time. It also takes work on a number of fronts. Here are three areas that, in my experience, it is particularly fruitful to focus on:

- establishing common ground;
- establishing common language;
- understanding and appreciating difference.

In all three areas, storytelling rooted in purpose can be an incredibly potent tool. Let's consider each in turn.

Common Ground

When the team at CMS conducted the first round of Open Door listening sessions in the run-up to merger, there were two particular themes that emerged very clearly from every single session: first, that everyone present highly valued the friendly, low-hierarchy, intimate culture that they were currently working in, and secondly that everyone present was concerned that this culture would be challenged in the new firm. Once this theme had emerged loud and clear from several groups in all three firms, the team was able to start playing it back, and this proved to be very helpful – people liked hearing that everyone wanted the same thing; and the gentle irony of everyone sharing the same fears about 'the others' helped to dispel those fears.

Annette Simmons says:

Narration simultaneously chooses and communicates a particular point of view.[18]

She reminds us that it's all about perspective – that the story of the three little pigs, for example, is a very different story depending on which pig you are (or if you are the wolf!). So when, as in this example, you want someone, or a group of someones, to see or understand something they are not seeing, a story can help to show them someone else's perspective. It was persuasive and powerful for me to be able to tell people about standing in a room *just like this one*, full of people *just like them*, in front of a flipchart *just like this one*, having a conversation *just like this one*.

Common Language

Another way in which stories can help to build trust between people is by giving them a common language. Tony Angel, who led the magic circle law firm Linklaters from 1998 to 2008 during a decade of profound transformation and subsequent success, with hindsight now points to the decision to send all partners on a Leading Professional Services Firms programme at Harvard as being perhaps one of the most important factors in driving through important changes that needed to happen in the business at that time. Not because the partners necessarily learned anything radically new during the programme – though the content was excellent and challenging, and quite radical in terms of the insights it offered law firms at that time – but because the common experience of having participated in the programme and, in particular, the case study methodology which is basically just stories – gave the whole partnership a new common language with which to discuss the challenges and opportunities in their own firm. It brought a degree of objectivity and distance and enabled partners to discuss issues which were very emotionally loaded with both courage and care.

This experience chimes with the following brilliant quote from Bruner, which reminds us that stories do not just passively describe the world; told and shared, they help us to shape it:

Through narrative, we construct, reconstruct, in some ways reinvent yesterday and tomorrow. Memory and imagination fuse in the process.[19]

Understanding and Appreciating Difference

Sometimes, the route to deepening trust is not so much about moving people onto common ground, but helping them to reach out across a divide which will remain. Often, such divides can be a good thing – the two sides bring different perspectives and skills and experiences – and the trick is to be able to speak across and into the gap to find congruence and mutual appreciation, even when the facts seem diametrically opposed. Stories can handle paradox and nuance like this well – think for example about the story I told at the beginning of this book about my first year as a mother. You did not conclude having read that story that that first year was unremittingly dreadful and I hated it; nor did you conclude that I loved every moment. In truth, you probably did not conclude anything in particular, or feel that you were expected to – you just let the story sit there and do its work; a little bundle of insight and truth that doesn't need every crease ironed out.

Once an organisation achieves this level of deep trust, where people are on common ground and speak a common language and yet deeply understanding of and appreciative of the differences between them, the ability to communicate at a level that goes beyond clarity to enlightenment is hugely enhanced – because people are in a frame of mind that will ensure they can listen and understand, and because the people doing the communicating will understand how to reach their audience in their choices of both perspective and language.

Stumbling Blocks

Just as there are many things that can make it hard for us to be good listeners, there are many things that can present a challenge to our effectiveness as communicators. This is true whether we are seeking primarily to bring clarity or to ignite a deeper meeting of minds. We have already talked about sheer volume and overwhelm, and about timeliness, and we have considered whether the best person is communicating the message. Let us consider a few other common stumbling blocks and how best to deal with them:

- **Making it all about you – your desire to be right, your biases and assumptions** – this is a tough one, because if you've a fire in your belly and are weary of keeping it in, there's a good chance you feel pretty invested in the message you are communicating, know your stuff and have thought about the topic in question longer, harder and better than pretty much anyone else in the room. Too bad. You need to hold your message lightly. Remember, you are *not* the message – you are the medium through which the message flows. Nobody is obligated to listen to you – you have to make them want to. Nobody is going to be anything other than who they are – you have to take them as you find them. The single most effective thing you can do as a communicator, whether your aim is simply to transmit information effectively or to effect a fundamental shift in attitudes, is to follow all of the same guidelines as apply when you are listening – put yourself squarely in the other person's shoes, put their thoughts and feelings before your own, remain open and receptive, and do not judge or criticise your audience. Storytelling can be a very useful device here to show your audience that you understand the world from their perspective. This will help to build trust and enable them to be more open and receptive to your message.

- **Making it all about them – people pleasing and conflict avoiding** – having said all of the above, you nevertheless have a message to convey. Perhaps not everyone will like the message. Perhaps it will lead to conflict and some awkward conversations or tough decisions. Convey the message anyway. Do it with as deep an awareness of your listeners' perspective as you can, and do it with the utmost regard for their feelings, but do it. The lack of clarity and, ultimately, the lack of trust that inevitably follows from any attempt to fudge or dodge a difficult message is never worth it.

- **Hyperbole, and an excess of positivity (or negativity)** . . . as in 'I am so unbelievably excited to announce that we have won this incredible deal which is super ground-breaking for the most spectacular client in the universe'. This is problematic in two ways. Firstly, just in terms of how language works and how we interact with it, overuse of hyperbole makes listeners switch off. It limits your dynamic range as a communicator and makes it harder for listeners to discern the nuance in the message. IT'S LIKE WRITING ALL YOUR EMAILS IN CAPITAL LETTERS. The other problem, touched on earlier, is that while positivity and confidence from leaders is vital, excessive positivity can undermine credibility if the lived experience on the ground is very different.

Strike the right note, though, and the effect can be transformational: early one morning in the teeth of the hardest few weeks post-merger at CMS, the senior leadership team sent an email acknowledging that things were difficult, explaining how they were addressing the biggest issues, thanking everyone for their persistence and grit, and making themselves available to chat. When this email – full of optimism but seasoned with reality, humility and empathy – hit inboxes, I swear you could hear the hiss as the pressure eased. Nothing objective changed that morning, but a particularly well-judged piece of communication created goodwill and galvanised people for just a little longer until things could turn a corner.

Notes

1. Einstein, Albert, 'The World As I See It', *Forum and Century*, 84: 193–194, 1931.
2. Lencioni, Patrick. *The Advantage*, Jossey-Bass, 2012.
3. The almost certainly apocryphal result of the relay by military radio of the command 'Send reinforcements, we're going to advance' – see https://quoteinvestigator.com/2011/08/26/reinforcements/.
4. 'We have two ears and one mouth so that we can listen twice as much as we speak'. Epictetus, 55AD.
5. '. . . So don't forget it, it's just a silly phase I'm going through . . .' With thanks to 10cc.
6. Argyle, Michael, *Bodily Communication*, 2nd edn, Routledge, 1988.
7. Pickering, Marisue, 'Communication' in *Explorations: A Journal of Research of the University of Maine*, 3(1), 16–19, 1986.
8. Rock, David. *Your Brain at Work*, HarperBusiness, HarperCollins, 2009.
9. Bell, Rob, *Something to Say*, Online communications seminar programme, @ https://robbell.com/portfolio/something-to-say/ downloaded and listened to in August 2017.
10. I am guilty of this myself. I recently joined the mailing list for the Women's Equality Party. Emails arrive on a regular basis, but I must confess I don't read all of them, partly because I'm sometimes distracted by fighting my own small-scale feminist revolution on several fronts. But this morning, I got an email from Sandi Toksvig, who is one of the founders of the party, and my inner voice went, 'Ooh, Sandi Toksvig', and I read the email.

11. Valla, Clement, *A Sequence of Lines Traced By Five Hundred Individuals*, Vimeo Video, posted in 2011, viewed in July 2018 at https://vimeo.com/18998570.

12. Bell, Rob, *Something to Say*, Online communications seminar programme, @ https://robbell.com/portfolio/something-to-say/ downloaded and listened to in August 2017.

13. Lencioni, Patrick. *The Advantage*, Jossey-Bass, 2012.

14. Lencioni then gets pretty didactic. An organisation's core purpose must, he says, *must* be completely idealistic, and the purpose *must* have something to do with making people's lives better. As will be apparent from my riffs in Chapter 1 on the various conceptions of the concept of purpose, I disagree with Lencioni here. I think different leaders can express a purpose which is personal to them or to their division of an organisation, provided it is consistent with, and essentially a strand of, an overarching corporate purpose. In practice, purpose almost always will be something idealistic, because for most of us it is something aspirational and just outside our grasp that most ignites us – neurologically, emotionally – gets us into flow, and keeps us engaged. But I don't load that with any normative or moral 'must'. Ditto the notion that a purpose generally entails making people's lives better. Experience suggests that that is almost always true, but it is not, in my mind, a necessary characteristic of purpose.

15. Scarlett, Hilary, *Neuroscience for Organizational Change: An Evidence-Based Practical Guide to Managing Change*, Kogan Page, 2016, and in discussion at the IoD, London in June 2016.

16. Heffernan, Margaret, *Beyond Measure: The Big Impact Of Small Changes*, TED Books, Simon & Schuster, 2015.

17. Meyer, Erin. *The Culture Map – Decoding How People Think, Lead and Get Things Done Across Cultures*, PublicAffairs, 2014.

18. Simmons, Annette. *The Story Factor: Inspiration, Influence and Persuasion Through The Art Of Storytelling*, Perseus Publishing, 2001.

19. Bruner, Jerome, *Making Stories: Law, Literature, Life*, Harvard University Press, 2003.

Chapter 14

Simplicity

To allow oneself to be carried away by a multitude
of conflicting concerns, to surrender to too many
demands, to commit oneself to too many projects, to want
to help everyone in everything, is to succumb to the
violence of our times.

Thomas Merton[1]

There is a town in The Netherlands that has no traffic lights. Twenty years ago, Drachten, home to 50,000 people, was a pioneer town in a project known as Shared Space, the brainchild of a traffic planner, Hans Monderman, in which all traffic lights and other signs, road markings and 'street furniture' were removed. Since then, cars, cyclists and pedestrians have indeed shared their space, road deaths have reduced dramatically, and tailbacks and congestion are almost non-existent. It is a model which has since been replicated, or at least experimented with, in other towns and cities across the world, and while it is not without its challenges and detractors, the results are startling and contain important lessons for us that extend far beyond traffic planning.

Look at what Monderman said when interviewed in *The Daily Telegraph*, seven years into the project:[2]

'We want small accidents, in order to prevent serious ones in which people get hurt,' he said . . .

'It works well because it is dangerous, which is exactly what we want. But it shifts the emphasis away from the Government taking the risk, to the driver being responsible for his or her own risk.'

Monderman then goes on to compare his traffic planning philosophy to the way an ice rink works:

Skaters work out things for themselves and it works wonderfully well. I am not an anarchist, but I don't like rules which are ineffective, and street furniture tells people how to behave.

This short interview contains, in a nutshell, pretty much everything I want to talk about in this chapter. I want to talk about simplicity. We live in a complex and dynamic world, and it is easy to conclude – and many do – that the way to address that complexity and to minimise uncertainty is to add more stuff – more structure, more processes, more rules. This is particularly true at times of change, and particularly true at scale. As Edward De Bono says,

There is no natural evolution towards simplicity.[3]

But in this chapter, I want to make the case for the opposite approach. I want to challenge our attitudes to risk, responsibility and rules. In life as in Drachten's town centre, it's dangerous *because it is meant to be*. We should create organisations full of drivers who take responsibility; ice-skaters who work things out for themselves.

Uh-oh! (More!) Mud

The problem is, things don't *feel* simple. They feel devilish complicated and there is a lot to do. In May 2017, the month the CMS merger went live, at a time where I needed to spend the least possible amount of time at my desk and the largest possible amount of time walking around talking to people, I was receiving on average well over 1000 emails a day. I am by no means alone in this – plenty of people across all manner of sectors tell me that their email traffic regularly (*regularly!*) exceeds this number. And this is not a problem only of volume – the substance of emails is often complex and needs to be teased out, and the different perspectives, interdependencies and nuances grappled with.

In their 2014 Global Human Capital Trends report, Deloitte reported on the 'overwhelmed employee' and made the case for a radical simplification of the work environment.[4] Their research found that information overload and the always-connected 24/7 work environment are reducing productivity and engagement. The report calls on organisations and leaders to do more to help people to manage their work in this context – to help them create more time to think and produce; to create smaller, more agile teams; to manage expectations; and to simplify practices and systems. This is not mere benevolence: in hard-nosed commercial terms, people who feel hidebound by structures, rules and processes will be less engaged, less productive, less creative and less accountable than those who do not.[5] This means that if we yield to the temptation to achieve short-term control and (the illusion of) certainty via processes and rules and structures, we are doing so at the cost of longer-term value creation.

To be clear, this does not mean that we should be simplistic. There is complexity and there is nuance, and these are important. One way of thinking about the difference between simplicity and simplistic (or simplistic-ity?!) is that simplicity is *post*-complexity; it can only be achieved by wading through complexity (Uh-oh, mud! Thick, oozy mud . . .) and emerging on the other side. Simplistic-ity by contrast assumes no complexity, and so doesn't even venture in.[6] My argument is that we need to do the work, grapple with complexity and then make things as simple as we possibly can, including by shedding (or resisting the temptation to create in the first place) anything extraneous by way of bureaucracy or distraction. Netflix describes this as eliminating barnacles. In this way, we will both increase our efficiency, and achieve a sophisticated simplicity[7] – and so will be able to focus our intelligence, attention and energy on the fine details and nuances that truly matter. Being purpose-led and telling compelling stories will help us immeasurably as we try to achieve this.

Let's consider how – and why – to achieve maximum simplicity in seven areas in turn:

- **simplicity of message**
- **simplicity of information**
- **simplicity of decision making**
- **simplicity of execution**
- **simplicity of structure**

- **simplicity of rules**
- **simplicity of expectation**

Simplicity of Message

Grant Thornton exists to 'create a vibrant economy', while Deloitte aims to 'make an impact that matters'. Unilever seeks to 'make sustainable living commonplace' and BT Group aims 'to use the power of communications to make a better world'. Franke, with Christophe Stettler's help, discovered that their highest purpose is to 'Make it wonderful', and perhaps best of all, Konditor & Cook exists to 'spread joy through cake'.

Organisations that have done the work and discovered, distilled and articulated their purpose have an advantage when it comes to being able to convey a message to their people, their customers and the market. They have a way of expressing who they are and what they are about that is simple, memorable and capable of repetition. The repetition bit is good news – remember David Rock's third condition for effecting changes in behaviour – we need to give people time to deeply understand and accept the message. We need to give them a reason to keep walking that new neural pathway again and again.

Lencioni says that great leaders understand that one of their main roles is to be Chief Reminding Officer[8] – a build on Jack Welch's statement that the role of a CEO is to be Chief Meaning Officer – to tell people where an organisation is headed, why, and what it in it for them. We should not, Lencioni says, confuse the mere transfer of information with the audience's ability to understand, internalise and embrace the message being communicated. Embracing only happens if you hear a message over time, in a number of different contexts and from a number of different people. What sorts of messages are susceptible to being communicated in this way? Not a series of facts, like some sort of Kim's Game, but stories.

Years ago, I left the practice of law to go and work as the bag-carrier for an incredibly driven and passionate managing partner who was going to transform his firm if it killed him. One of the things I was most struck by in the first year or so working for him was that he told the same story about his vision for the business over and over and over and over again – in London, Frankfurt, Beijing, New York . . . to 500 people or half a dozen people . . . at breakfast, at dinner . . . and each time, with as much conviction, detail and freshness as if it was the first time he had ever told it. I couldn't get my head around it. This brilliant, restless intellect, throwing off ten new ideas every minute, and yet he had the patience and discipline to repeat, repeat, repeat. It is a trait I have seen replicated in almost every great leader I have known or observed since. In *Turn The Ship Around!*, David Marquet, former captain of the USS *Santa Fe*, also talks about the need for relentless, consistent repetition. He says:

> When you bring in something new, something that has never been seen before . . . [the crew] hear and think they know what you mean, but they don't . . . They can't see in their mind's eye how it works.[9]

And so, he says, you need to repeat the same messages over and over. Leaders generally vastly overestimate how aware the people in their organisation are as to its purpose and priorities. Leaders need to dramatically, drastically overcommunicate, and be crystal clear and boringly consistent in the messages they deliver.

In *The Culture Code*, Daniel Coyne explores how two very different organisations – Danny Meyer's highly successful restaurant chain, and Pixar Studios – use aphorisms and pithy catchphrases – *Read the guest. Put us out of business with your generosity. Face towards the problems. Fail early, fail often.* – to encapsulate key stories and tenets and enable them to be remembered and embodied and to cascade through the organisation in order to guide behaviour. These are vivid decision-making heuristics.

Just a note of caution here, though. Sometimes, the whole concept of organisational purpose is met with a degree of cynicism and a particularly pithy statement of purpose is perceived as 'just a strapline', a marketing ploy. Crafting a perfect sentence is not the point. There is a need for clarity, yes, but there is also a need for nuance and inclusion. At CMS, our purpose statement is a little longer than 'spreading joy through cake', but it works for us.

The simplistic/simplicity distinction is relevant here – there is a world of difference between having truly done the work to discover and articulate one's purpose and having crafted it into something succinct and memorable, and having come up with a catchy slogan on the back of an envelope one idle afternoon. In truth, though, that distinction is only apparent to those on the outside, and a statement of purpose will only truly be accepted as such, if there is integrity as between an organisation's words and their subsequent actions. A commitment to purpose is ongoing, every single day.

Simplicity of Information

In telling the merger story in Part Two, I touched on the paradox that the more you tell people sometimes, the less they feel they know. This might be in part because you ignite curiosity and lead people to wonder and to ask questions that they might not otherwise have asked. But it is also – I think predominantly – linked to Deloitte's 'overwhelmed employee' concept; if people do not know how to navigate, prioritise or curate the vast amount of information that is available to them, they will be left still feeling ignorant, disengaged and isolated. The default is to give your attention to whatever is shouting loudest; the red light that is blinking most furiously. In a 21st-century take on the fable about the boy who cried wolf, people vie for each other's attention by sending emails with headers saying, 'PLEASE READ' in block capitals, and then 'ACTION REQUIRED', and then 'OKAY, I REALLY REALLY MEAN IT THIS TIME'.

There is another, more insidious, risk of simply sharing facts – as if people feeling ignorant, disengaged and isolated were not danger enough – namely that, as an organisation, you cede control of the message. As Annette Simmons points out,[10] facts are no sort of method of influence at all. Facts are neutral until people ascribe meaning to them. And the meaning they ascribe depends on the story they are telling themselves. If an organisation just carries on blithely pumping out facts with no regard to the narrative into which they are being incorporated, it should come as no surprise if what it hears being played back is an odd distortion of the intended story. Organisations and leaders need to give people the story first; the framework and context into which facts can then be absorbed.

At CMS, the team aimed to be very considered in its approach and to help people with their attention, prioritisation and curation by putting a narrative arc around all of the information that they needed to share. The team did this at the macro-level at the outset of the whole integration process by storyboarding everything, splitting the period ahead into distinct seasons and weaving the key facts that they knew they needed to communicate into different parts of the story. The team then shared this idea of seasons across the firm, and helped people to contextualise by constantly looping back to that big picture. They tried to use stories, in other words, to cut through the complexity and chaos and help people both to access and use the information they needed, and to avoid feelings of overwhelm and isolation.

In my experience, this whole challenge around giving people the information they need to do their job and feel engaged and included, without overwhelming them, is one of the most profound challenges facing businesses today, and the added uncertainty that change brings makes this all the more profoundly difficult. Stories have a critical role to play in tackling this, and enabling people to feel well-equipped to do their work.

Simplicity of Decision Making

This brings us to decision making. The received wisdom is that the average adult makes around 35,000 remotely conscious decisions every single day. For senior leaders, it seems likely to be more, notwithstanding that Reed Hastings of Netflix prides himself on making as few decisions as possible[11] – a neat KPI for a purpose-led, networked and agile organisation that has learned how to empower locally and devolve decision making accordingly.

In relation to those business decisions that do have to be made, however, purpose is both a useful and vital lodestar. Useful, because it provides a shortcut to good decisions and confidence in the conclusions reached without starting from first principles every time amidst a sea of options, and because it helps with prioritisation – there are some decisions which organisations can easily spend hours on which simply do not matter and do not advance the purpose. And vital because if purpose is not used to guide decisions in this way, it is not truly purpose, and the strapline-cynicism noted above is justified. As Aristotle said:

We are what we repeatedly do. Excellence, then, is not an act but a habit

Make It Easy on Yourself

We need to get into the habit of making decisions in line with our purpose. This is true from the smallest and simplest of decisions – how you will greet colleagues first thing in the morning – through to the biggest and most complex – whether or not to merge your law firm with two others. Laying down a habit is like walking a desire path through the long grasses of the individual brain, and the corporate mind – it takes work, but once established it eases the cognitive load by making decision making easier. If you have established a clear habit of making decisions aligned with purpose, it becomes the default position over time.

Chip and Dan Heath, in their book *Switch*,[12] draw on the metaphor of a small rider atop a big elephant walking along a path and argue that in order to effect change, you have to do three things – direct the rider, motivate the elephant and shape the path. Purpose can help with all three of these. You can direct the rider by reinforcing the purpose, telling stories and making it clear, therefore, what the end goal is. You can motivate the elephant by telling stories that resonate and making everyone feel part of something, building confidence and understanding. We have already considered both of these at length. But finally, you can shape the path by simplifying things so that the change is factually easier for people to achieve. You do this by tweaking the environment – clearing unnecessary systems and rules out of the way. You also shape the path by laying down new habits and by rallying the herd to all move together in a particular direction.

This is a key role of leaders and organisations in facilitating purpose-led decision making; you have to rig the environment for success by shaping a clear path – through the processes and systems and rules you have or don't have (see below), through role-modelling and storytelling, perhaps through catchy aphorisms of the sort discussed above, and perhaps also by considering creating a framework for decision making that is in line with your purpose.

It's Complex

It is easy to see how this works for simpler decisions, but many decisions are much more complex. How should we approach making complex decisions, particularly at times of change, in the simplest way possible? And how can we be sure that we make such decisions in line with our purpose?

First, let's consider what we mean by complexity. Margaret Heffernan makes a neat distinction between 'complicated' and 'complex'.[13] Complicated situations, she says, are the familiar problems of the industrial age – there may be many aspects to a particular situation, but it is essentially a linear problem, and can be broken down into stages which can be tackled, in siloes, sequentially, and a solution or decision reached at the end of the process. Complex situations, by comparison, a defining feature of the information age and VUCA times we live in, are multi-faceted, with many factors which are interdependent. They look more like a spider's web than a straight line, and cannot be tackled in a sequential, siloed way.

When we are faced with this sort of decision, a lack of data is rarely the problem. Quite the contrary; we are often drowning in information, and the first task is to discern what is relevant and useful. And yet, the more complex a situation, the more we tend to resort to analysis. This is perhaps a particular temptation in professional services organisations. Lawyers, engineers . . . we want to break the problem down into manageable chunks. We want to predict, control and measure everything. But this is simply impossible to do. And to make matters worse, chaos theory reminds us that, in a complex system, our very efforts to do this will themselves impact the situation and add to the unpredictability.

So what to do? First, you need to actively value simplicity, and be determined to seek it – rather than just being surprised and delighted if a simple solution happens to present itself. This may mean making trade-offs and choices, and possibly also discarding existing elements. In Dachten, a conscious choice was made to slightly increase the risk of small accidents, and to shift more responsibility onto drivers. It also means investing the time to create and consider different alternatives and possibilities. You can only do this if you have all the expertise and mastery you need in the room – hence the importance of diverse teams.[14]

We then need, to some extent, to surrender – to 'be more Dachten'. This is scary – Dan Allender[15] quotes a pastor who describes it as like juggling flaming chain saws. But we do not have to – in fact we *must not* – juggle alone. This sort of decision making and problem solving demands inclusion in the broadest sense, and a collective approach to leadership – you need people with different perspectives, background and areas of expertise around the table – real or virtual – together, all of them with a voice. Purpose helps immeasurably when this is happening – it gives everyone the common ground they need to stand on, and a framework within which to make a decision.

Simplicity of Execution

Once a decision has been made, we need to make something actually happen. This can often be the hardest thing in any organisation, and particularly in large, flat, complex organisations, at times of change, when people do not know each other well and it is not clear who is responsible for what, or who has authority. Again, it can be tempting to try to deal with this by imposing more structure, more meetings, more rules, more processes, more lists.

But this way lies bureaucracy, stagnation and misery. Much better, as we discussed in Chapter 12 on Agility, to make things happen by engaging and empower everyone in the organisation to act in accordance with the purpose. Chapter 12 discusses at length how to do that, including the role that stories can play, and there is no need to repeat those ideas here. But just one further thought, with a focus on simplicity. In *Making Ideas Happen*, Scott Belsky talks about how creative journeys all begin with a spark in one person's mind.[16] From there, the challenge is in figuring out how to get others to understand and support the idea through to fruition, as if it were their own. The same applies to implementing any decision. Leadership is not, says Belsky, about making people do things – but rather about instilling,

> . . . a genuine desire in the hearts and minds of others to take ownership of their work on a project. Only then can we act together, motivated by a same purpose.

This model of leadership, though, requires leaders to:

- have a high level of trust in those around them to execute;
- to embrace decisions that they may not have made; and
- to be focused on outcomes rather than process.

This is easily said but can be challenging in practice – not least because many of the most passionate and idea-generating leaders have views on everything – they have a propensity to micro-manage and control and do believe that there is a right way to do things as well as right things to do. They have strong views on process! . . . But leaders who can get this right, as the leadership team at CMS did, can turbo-charge their organisations and get lots done, really well, in a remarkably short space of time.

Simplicity of Structure

Recently, I embarked on a piece of work with a new organisation, and asked a member of the Board to outline the organisational structure for me. She started with a pen at the top centre of the whiteboard and drew a box to represent the Board, and then drew a line to the Chairman. She drew the management team, various regional and service-line leaders, some technical experts, various sub-committees of the board. She drew the digital advisory panel . . . then she drew some dotted lines to show how information and ideas are sometimes shared, and how decisions are sometimes made . . . she stopped as she saw my eyes glaze over and we looked at the tangled web in front of us.

'It doesn't *feel* that complicated,' she offered . . .

In *Fast/Forward*, Birkinshaw and Ridderstråle draw on the world of complexity theory – which explains how languages are born and develop, how ants build colonies and how geese fly in v-formation – and advocate for organisations to become more self-organised and less hierarchical.[17] This, goes the theory, enables better decisions to be made more quickly in complex environments. We need to be willing to operate 'on the edge of chaos', and the primary role of leaders is to empower rather than control. In a similar vein, Julia Hobsbawm, in her book *Fully Connected*,[18] talks about 'connecting inwards' and casts around a little, as I am now, for a word or phrase to encapsulate what we are talking about here. This, she says:

> is something bigger than . . . internal communications, and it is bigger too than . . . learning and development.

It is indeed. We now have the technology to identify, track and illustrate – in beautiful pictures that look like airline maps or neural pathways – the social networks within organisations. It is fair to say that these social networks – who actually knows who, who we go to for help, who we share ideas with – bear little or no resemblance to any official organograms or power structures. Hence the scribbles on my whiteboard.

The leadership of an organisation fights this 'something bigger' at their peril. People are people – they talk to each other, connect, form bonds, and friendships and affiliations, and share information and ideas. Better by far to embrace this reality, to maximise its value to the organisation in terms of its capacity to share information, to generate ideas and make them fly, and to build engagement and belonging. Hobsbawm points to the unsavoury example of criminal and terrorist networks – highly devolved and informal and yet highly effective – often in marked contrast to the centralised, top-down and cumbersome organisations that try to take them on. Lencioni[19] counsels that one of the most effective ways of cascading information through an organisation is by spreading 'true rumours' – hearing consistent stories, contextualised and told by those immediately around you builds awareness, belief and buy-in much faster than a well-orchestrated, facts-based, electronically-delivered communications campaign.

This makes me think of the jaw-dropping wonder that is a murmuration of starlings. One bird in a flock reacts, say to the proximity of a predator, and immediately the whole flock responds as one – a bird thousands of birds away from the original instigator instantly knowing what to do such that the whole flock transforms into a single living cloud, twisting and swooping. It looks like magic, but we now know it's about attentiveness – akin to the deep listening we discussed in Chapter 13 – each bird paying absolute attention to the signals coming from the six or seven birds closest to it.[20] No formal organisational structure has a chance of replicating something so powerfully singular and responsive.

And yet, to be clear, for all its advantages, self-organisation is rarely – if ever – the entire answer. The first parts of that structure chart that my colleague drew were absolutely right and operate in a clear way – it is clearly important to know what decisions are of sufficient import to go to the Board, for example, or to know what it is the management committee's job to get on with doing. We know what matters are reserved to the senior and managing partner under the firm's articles of partnership, and which matters require a vote of all partners. This is important – some decisions need the broadest perspective and the broadest buy-in. It is important, too, to build clarity and trust in a robust process around the most material decisions – culturally, strategically or financially.

As to the myriad other ways in which people want to organise themselves to share information, brainstorm ideas or drive projects forward, self-organisation is a simple and effective way to proceed. But it is not without its risks. It depends heavily on an assumption of competence, good intent and intrinsic motivation on the part of the people working together. It also presumes a good process for permitting, catching and rectifying mistakes, and foregoes the utmost efficiency, allowing for some degree of duplication of effort.

This is where purpose and stories can help enormously. It is purpose and stories that will keep people connected within an organisation. It is purpose and stories that will enable them to navigate complexity and find the other people and the information that they need, when they need them. It is also purpose and stories that will keep them connected to what they do, owning it, understanding its value and worth to the organisation, and operating within the broad parameters that the common purpose lays out.

Simplicity of Rules

So, let's assume you have this self-organised group of people in your organisation, who are crystal clear on their message, super decisive, and absolutely sh*t-hot at getting things done. Don't you need some rules? What if they, I don't know, do something *wrong*?

In my experience, there are two distinct reactions when it comes to rules.[21] In my family, most of the children love a good rule, ideally at some-one else's expense. But I have one child who is positively allergic. Stand-ing behind the yellow line on the platform, not poking bees, not mooning at grandma . . . the very existence of a rule seems to evoke in him a deep desire to break it. I would say I don't know where he gets from were it not for the fact that nothing irritates me faster in a professional context than an apparently arbitrary rule. Forty-two-page policy governing the use of social media? Get a taxi home on the firm's account at 9.05pm in high-summer after a client drinks party, but not at 8.55pm in February when you've been working for 12 hours solid with a headcold? It's hard to find a rule that can't be paraphrased as 'Please. Just don't be a muppet.'

I'm deliberately overstating. Of course there are some rules which are absolutely vital, non-negotiable bright lines, and rightly so. Particularly in the context of a law firm, there are rules to protect clients, their confi-dentiality, the fair operation of the market, proper use of money, and so on, and all of these are critical. Ditto all the vital rules that protect people from abuses of power.

There is also a place for well-thought-through and consistently applied processes and policies, and this is particularly true at scale and where the need for fairness, consistency and transparency is high. So, for example, during the CMS merger, one member of the new HR team was tasked with writing up detailed process notes for every aspect of every HR process in order to ensure we could be consistent across the new firm.[22] The existence of those notes, and the consistency of process which resulted, was immensely valuable in terms of us being able to execute on all manner of things smoothly, and also in terms of building trust.

No. What I am railing against here is the proliferation of unnecessary rules and processes in the mistaken belief that this is an effective way to retain order and control as things change, get bigger and/or get complicated. When an organisation is spending more energy trying to avoid errors than achieving excellence, there is a problem.[23]

Laloux[24] argues that you only need rules if there is an underlying sense of fear – if you have people firmly rooted in purpose, you can take away rules and they can and will achieve more. President Dwight D. Eisenhower said:

Leadership is the art of getting someone else to do something you want done because he wants to do it.

In all seriousness, the most liberating moment in my career came when a mentor advised me:

It's much better to ask forgiveness than permission.

If you are clear on your purpose and your culture, the conversation and the tacit contract with your people around their behaviour becomes about what kind of organisation you want to be, not about rules. If you are clear about your purpose and values, you are free to improvise as the circumstances require, and to empower and enable your people to do likewise. You can duck and weave around obstacles, tack across choppy waters, zig-zag up a mountain – pick any metaphor you like, the point is the same – clarity about destination; scope to improvise as to route. This is liberating and opens up the opportunity for space and simplicity. You don't need lots of rules and policies and procedures and committees and boards and papers. You need parameters, for sure – an improvising member of a jazz band isn't playing preordained black notes on a white page under a conductor's hand, but she is not just playing anything she likes either (she really isn't!). But this is about being facilitative rather than prescriptive. Less bureaucracy and fewer rules; more freedom, creativity, discretionary effort and collaboration.

Simplicity of Expectation

In her wonderful book, *The Sweet Spot*, Christine Carter tells a story of how it can be hard to think straight in her household when all four of her children and their friends sing (different songs!) loudly.[25] Maybe this anecdote sticks with me because I can so deeply relate to it – indeed, as I write this I can hear someone in the background singing along with *Alvin and the Chipmunks*, and someone else channelling Sia . . .

Anyway. Carter talks about having found a degree of peace or at least equanimity in this situation by simultaneously accepting the imperfections of others as an intrinsic part of what makes them beautiful, and recognising that whatever is irritating us is transient and will not last forever. This is essentially the Japanese concept of *wabi sabi* – the aesthetic of finding beauty in a thing's flaws.

A purpose-led organisation can draw on this wisdom when it comes to its sense of self and its ambitions for, and expectations of, itself and its people. We are living in a time when, with some justification, public trust in institutions and organisations is low. People also have access to huge amounts of information, and are quick to be cynical and to point the finger at any apparent hypocrisy, any gap between words and actions. This is why there is a danger in the 'strapline' approach to purpose that we touched on earlier – it is not credible to simply impose a statement of purpose on the top and then carry on doing business exactly as you were before. Not only will that not yield benefits, it is likely actively erode trust and stakeholder value.

But where does that leave the truly purpose-led organisation? There is hope, provided there is a true commitment to purpose in the organisation, and a careful curation of the stories it tells. It is possible, and right, to have the highest ambitions and aspire to true excellence, true clarity of purpose, consistently lived and delivered; and yet to hold onto the truth that there will be mistakes and shortcomings and disappointments along the way. In the Christian tradition we call this grace; in the Jewish tradition, there is a wonderful and oft-instagrammed teaching in the Talmud which says, in essence:

It is not incumbent upon you to complete the work, but neither are you at liberty to desist from it[26]

Much as the phrase makes my toes curl, all purpose-led organisations are 'on a journey' and the stories they tell will be all about what they have learned on the road, from their mistakes and shortcomings as well as from their successes. Wabi sabi.

Notes

1. Merton, Thomas, *Conjectures of A Guilty Bystander*, Bantam Doubleday Dell Publishing Group Inc, new edn, 1994.
2. Millward, David, 'Is This The End Of The Road For Traffic Lights?', *The Daily Telegraph*, 4 November 2006.

3. De Bono, Edward, *Simplicity*, Penguin Life, 1999.
4. *Global Human Capital Trends 2014: Engaging the 21st Century Workforce*, Deloitte, 2014.
5. See for example: Ariely, Dan, *Payoff: The Hidden Logic That Shapes Our Motivations*, Ted Books, Simon & Schuster, 2016; Cable, Daniel, M., *Alive at Work: The Neuroscience of Helping Your People Love What They Do*, Harvard Business Review Press, 2018; Heffernan, Margaret, *Beyond Measure: The Big Impact Of Small Changes*, TED Books, Simon & Schuster, 2015; and Rock, David. *Your Brain at Work*, HarperBusiness, HarperCollins, 2009.
6. De Bono, Edward, *Simplicity*, Penguin Life, 2015.
7. Garcia, H., and Miralles, F., *Ikigai: The Japanese Secret to a Long and Happy Life*, Penguin, 2016.
8. Lencioni, Patrick. *The Advantage*, Jossey-Bass, 2012.
9. Marquet, L. David, *Turn the Ship Around! A True Story Of Turning Followers Into Leaders*, Penguin, 2015.
10. Simmons, Annette. *The Story Factor: Inspiration, Influence and Persuasion Through The Art Of Storytelling*, Perseus Publishing, 2001.
11. Quito, A., 'Netflix's CEO says there are months when he doesn't have to make a single decision', *Quartz at Work*, 19 April 2018, reporting on an interview of Reed Hastings by Chris Anderson, TED curator in Vancouver in April 2018.
12. Heath, Chip and Dan, *Switch – How To Change Things When Changing Is Hard*, Random House Business Books, 2011.
13. Heffernan, Margaret, *Beyond Measure: The Big Impact Of Small Changes*, TED Books, Simon & Schuster, 2015.
14. De Bono, Edward, *Simplicity*, Penguin Life, 1999.
15. Allender, Dan B., *Leading with a Limp*, 2006, WaterBrook Press.
16. Belsky, Scott, *Making Ideas Happen: Overcoming The Obstacles Between Vision And Reality*, Penguin, 2011.
17. Birkinshaw, Julian and Ridderstråle, Jonas, *Fast/Forward: Make Your Company Fit For The Future*, Stanford Business Books, 2017.
18. Hobsbawm, Julia. *Fully Connected – Surviving and Thriving in an Age of Overload*, Bloomsbury, 2017.
19. Lencioni, Patrick. *The Advantage*, Jossey-Bass, 2012.
20. Ballerini, M., Cabibbo, N., Candelier, RF., Cavagna, A., Cisbani, E Giardina, I., Lecomte, V., Orlandi, A., Parisi, G, Procaccini, A., Viale, M., and Zdravkovic, V., 'Interaction Ruling Animal Collective Behaviour Depends on Topological Rather Than Metric Distance: Evidence From A Field Study', *PNAS*, 105, pp 1232–37, 2008, as distilled and discussed in Coyne, Daniel, *The Culture Code: The Secrets Of Highly Successful Groups*.

21. There are also two distinct two kinds of people in this world. People who say things like 'there are two kinds of people in this world', and people who don't. ☺.
22. I am just deeply grateful that there are people in the world who can do this – who actually *enjoy* doing this – when it would reduce me tears inside half an hour.
23. Marquet, L. David, *Turn the Ship Around! A True Story Of Turning Followers Into Leaders*, Penguin, 2015.
24. Laloux, Frederic, *Reinventing Organizations: A Guide to Creating Organizations Inspired by the Next Stage of Human Consciousness*, Nelson Parker, 2014.
25. Carter, Christine. *The Sweet Spot*, Ballantine Books, 2015.
26. Avot 2:21 in Pirkei Avot, attributed to Rabbi Tarfon.

Chapter 15

Energy

*All sorrows can be borne if you put them into
a story or tell a story about them.*

Isak Dinesen[1]

*The antidote to exhaustion isn't rest.
It's wholeheartedness.*

David Whyte[2]

T hings are always intense in the early days of a major change project. More than once during the early weeks of a project, I have cupped someone's chin in my hands like their mother and suggested they go home, eat something and then sleep. I am always struck by the sheer weight of personal responsibility that I feel at such times, sitting like a dull cold stone on the top of my chest. I feel over-caffeinated. Brittle. Such times require a lot of everyone in the business. The leadership team, the operational teams, the people on the ground, are all coming to terms with change, and new people and systems and environments.

These acute early-days situations have a lot to teach us about energy and resilience, their role in any change process, and about how our approach to change can either bolster or diminish individual and collective energy levels. That is what this chapter – the final chapter in this part of the book, the 'E' in our BECAUSE – is about.

Energy, or Resilience?

By Energy, I mean Resilience, right? Going back to our buzzword bingo, Resilience should be right up there next to Agility. It is a widely used term in organisations at the moment, and this is largely a good thing – being indicative, I think, of both a recognition of the turbulence that is an inevitable part of doing business in an uncertain and complex world, and a pronounced shift in organisations over the past five years or so towards taking wellbeing in general and mental health in particular much more seriously.

I am writing this chapter in a cottage on the North Cornish coast, marking the end of a heatwave by watching the wind whipping torrential rain across the bay and onto our little garden. At the foot of the garden is a little line of trees, which in another couple of decades may afford some degree of protection. But for now the garden is taking a beating. The fact that it nevertheless manages to be a pretty garden is because of resilience: the plants that the owners have planted here have been chosen precisely for their ability to withstand wind, salt, horizontal rain and narrow sandy soil. Hebes, escallonias, sea buckthorn, ceanothus, lavender, rosemary – tough, woody, hairy or hunkered down, they are all survivors. The word resilience comes from the Latin *resilere* meaning to rebound, or to bounce back, after being bent, compressed or stretched. When we talk about people and organisations being resilient in the face of turbulence or change, what we are essentially saying is that they have what it takes to face down the whipping salt wind and bounce back. Part of what we will consider in this chapter is how to help people and organisations to develop this ability and, in particular, how purpose and stories can help with this.

And yet, you will note that I have not succumbed to the buzzword. This chapter is not called 'Resilience'. That is not (only) because Resilience does not begin with E, but because my ambition and hope is higher than this. On the brow of the hill on the other side of our rainy Cornish cove, currently barely visible through the low cloud, are three towering wind turbines. These big girls are doing way more than simply being resilient. They are not simply standing there, facing down the wind and waiting for it to pass; they are actively harnessing it to create electricity. They are taking energy from the wind and using it to make something new. This is what we want for our people and organisations during times of change. In merging three firms to create one new one, the team at CMS were aiming to do much more than just survive – they were aiming to create a new organisation with more energy, more momentum, which would be more than the sum of its parts. By extension, they were aiming to create a context within which their people would flourish, not just cling on by their fingernails in narrow soil under torrential rain. And – most ambitious of all – the aspiration was that they should achieve that flourishing not just as an end goal once the merger was somehow 'done' and everything was as calm as a millpond, but *while the change was ongoing* – because, as we have discussed, change is the new normal. We need the change process itself to bring energy, to ignite the imagination, to incite discretionary effort, and to invite wholehearted participation.

And so in this chapter, I want to talk about Energy. Energy encompasses resilience and survival, and we will consider both of these in depth, but Energy is a bigger concept than resilience alone, and also includes inspiration, creativity, wholeheartedness and momentum, all of which we will also explore.

Why Does Energy Matter?

In making the case for caring about the energy and resilience of your people and your organisation, perhaps the most compelling place to start is by imagining a situation in which both were significantly depleted. If an organisation is without dynamic, creative energy, it is unlikely to be able proactively to spot opportunities and capitalise on them. If even its resilience is low, it may find that even reacting adequately to the vagaries of the market is a challenge. In these VUCA times, it will be the organisational equivalent of a hothouse flower in a Cornish rainstorm and, in all likelihood, not long for this world.

Within organisations, individuals who are lacking in resilience, are likely to be suffering, particularly during a period of significant change. In addition, they will be presenting the organisation with a real cost, in terms of increased sick days, higher attrition rates, poorer quality work and relationships, and reduced productivity and efficiency. On the upside, these signs and symptoms tend to be relatively easy to spot, and there are numerous options available to organisations to support individuals once an issue is known about – for example, by seeking to build engagement and by coaching them to develop some of the key skills and characteristics associated with resilience, as listed below.

The less visible, and so perhaps more invidious, risk for organisations relates to individuals who are 'resilient enough' to withstand change without any obvious dip in performance, but who have not been swept up in the vision of the organisation such that they are actively energised and wholehearted. In this scenario, organisations will find themselves lacking the firepower and creativity that comes when people are intrinsically motivated and invest discretionary effort in pursuit of an organisation's purpose. The previous chapters in this book are littered with examples of positive actions in pursuit of successful change which actively depend upon engaged and energised people showing up and participating – coalface mastery in evolutionary strategy setting, building community and networks, innovating, localised decision making, supporting one another . . .

Over time, if efforts to do these things flounder due to a lack of participation, more of the people who were previously engaged 'just about enough' will disengage, and this will slowly become apparent over time. As in a plane when the engines cut out, performance will drift slowly downwards, quality and efficiency decline, new opportunities be missed. There may be no immediate drama, but the trajectory is inexorably downwards. If, however, an organisation prioritises empowerment and engagement, the prize is an organisation full of resilient, wholehearted and energised people . . .

Imagine a community of people, guided by their purpose, and bonded by common experiences of both adversity and success, with a deep understanding of one another, standing on the cairn of their past achievements with the confidence to face new challenges, well equipped for an environment of constant change because they know how to let their strategy evolve on the ground and how to harness their collective intelligence to respond with agility to new opportunities by creating new things. People who thrive on the autonomy that comes from being free to make decisions, having simple systems which enable them, and as few rules as possible, but who are self-aware and have the self-mastery such that they always hold themselves and each other to account and aim for excellence.

All of this depends on energy, and all of it creates energy – we are standing at the entry point to a virtuous circle. That, right there, is the case for caring about resilience and energy.

How Does Change Challenge Energy and Resilience?

A prolonged period of profound change can become a perfect crucible in which to burn off energy and resilience. For those most directly involved in actually effecting the change, there is as a starting point the sheer hard graft involved. My own 'Bloody Mary moment', for example, came after fifteen months of very hard work. I was tired. There was a huge amount of work to do, and tackling a bottomless task list, always recalibrating and juggling priorities, can present a challenge to our energy and resilience. In any major change programme, many are to some extent also grappling with a degree of grief associated with what they are leaving behind as a result of the change, and the sheer amount of change that people are dealing with at once means that there are inevitably oversights, obstacles and setbacks which make a challenging situation tougher still.

More generally, consider Rock's SCARF model again.[3] People involved in change projects are often dealing with a huge amount of uncertainty, which presents their brains with a big challenge around the 'C for Certainty' part of Rock's acronym. There is uncertainty over the big things – *What does the future look like in this new firm?* – and the small things – *How do I send a bill out?* When I went back through all my archived emails from the CMS merger in preparing to write this book, I sorted them alphabetically by their original subject line to look for themes. There were lots of emails under C, N and O, of course. Lots under U for Urgent, and lots under H for Help. But, dwarfing all of these lists, there were hundreds and hundreds of emails listed under N for New. Up to a point, our brains like novelty – the anticipation of something new and different can trigger a dopamine hit in anticipation of reward – but the neurological work involved in dealing with lots of new things at once and laying down new neural pathways to map lots of uncharted territory, can be taxing and depleting.

As discussed earlier in this book, during periods of change, many people also perceive challenges to their 'S for Status', around their reputation and standing, and some perceive challenges to their Autonomy, particularly when it comes to feeling thwarted by new systems and processes. Relatedness takes time to build as it depends on personal relationships and belonging, and Fairness too depends on trust, a degree of transparency around decision making and an ability to navigate information.

All the research shows that, in neurochemical terms, our brains respond to the sorts of perceived 'social' threats which can arise in the context of organisational change in exactly the same way as they would respond to a perceived physical threat. It really does feel as though you are being chased by a tiger! When people are in a threat state, adrenalin and cortisol levels are elevated and dopamine levels are depressed as a result. Even over the short term, stress and elevated cortisol inhibits our ability to think clearly, increases negativity and depletes our self-control and our ability to manage our emotions. Longer term, studies have established a link between psychological stress and cardiovascular disease,[4] a depressed immune system and other physical conditions. Stress hormones also have a longer-term negative impact on the pre-frontal cortex, and on the hippocampus, which is responsible for memory creation and recall. Meanwhile, reduced dopamine levels lower mental performance, inhibit creative thinking and reduce our desire to do anything significant.[5]

It is clear, then, that periods of change present a number of challenges to an individual's – and by extension an organisation's – energy and resilience, and that these challenges can lead to potentially damaging consequences for the short and long term. However, these potential consequences are by no means inevitable and there are practical steps that individuals and organisations can take to boost their energy and resilience so that they can withstand, and ultimately even thrive on, periods of change.

The Essential Rhythm

So, how can you tell if your organisation and your people are resilient and energised and equipped with what they need to withstand the buffeting that change brings? In a nutshell, by virtue of something of that 'carrot' vision outlined above being apparent. There may also be some concrete KPIs in your organisation that you can track in order to look for trends – hours billed, NPS or other engagement metrics, attendance at events, participation in CSR activities, or surveys or crowd-sourcing activities, frequency and duration of sickness days, incidence of seeking support for stress-related issues, attrition and so on. Beyond that, you will need to watch and listen to your people . . .

Do people have a sense of belonging to the organisation? Are people generally positive, flexible and open in their approach? Do people readily step up and take responsibility? Is there a sense of intrinsic motivation? An innate confidence and the willingness to step out and take a risk? An ability to learn, reflect and re-engage? Resilient and energised people have a certain constancy, a rhythm, that seems to be fundamental. The poet Seamus Heaney said:[6]

Getting started, keeping going, getting started again—it seems to me this is the essential rhythm not only of achievement but of survival, the ground of convinced action, the basis of self-esteem and the guarantee of credibility in your lives . . .

At an organisational level, in a resilient and energised organisation some of the same traits will be reflected. The organisation will be remarkable for its positivity, candour and flexibility. It will have a notable confidence about it in the market and vis-à-vis its customers and its people, and will be willing to do things differently, to take a calculated risk. There will also be a sense of humility and a willingness to learn throughout the organisation, and a sense of ease rather than frenzy – a willingness to take the time to address the things that matter, a sense of being on the front foot, rather than just constantly reacting as issues arise.

Unexpected issues inevitably crop up. In a less resilient organisation, these issues can lead to a besieged, 'Life is one damn thing after another' type of stance, and a desire to simply address the issue as quickly as possible while pointing the finger of blame clearly away from oneself, and then move on. A more resilient and energised organisation can take a broader view, prioritise people and relationships, communicate expansively about what is going on and how, together, we are working to resolve the issue, and offer practical support and help. Energised organisations don't just withstand the buffeting of the unexpected setback; they use it as an opportunity to learn, connect and deepen relationships.

How Can Organisations Build Resilience and Energy?

Building Resilience

There are steps that organisations can take to help their people to build resilience. There is clear evidence, for example, that giving people a sense of control materially reduces cortisol,[7] and so the various actions that we have discussed previously around sharing information and otherwise reducing uncertainty wherever possible ('We don't know, but we'll know on Tuesday when we'll know . . .' etc), encouraging local decision making and reducing bureaucracy will all help in this regard. It can also be helpful to set short-term goals that people can readily achieve. This has the dual benefit of conferring control, and so reducing cortisol, and of generating dopamine by virtue of a goal having been achieved. Reminding people of their past achievements – helping them to stand on their cairn – can also increase dopamine, as can being liberal with praise and recognition, and reporting in real time on milestones and wins along the way, both things that the team at CMS did a lot of in the early days of the merged firm: daily thank you and praise notes to people, emails on *x pitches won; y directory listings gained*, etc.

Jane Clarke and John Nicholson, directors of the organisational psychology consultancy Nicholson McBride, have researched resilience extensively, and report in their book *Resilience* the findings of a study which included in-depth interviews with twenty-six people who had been identified as being particularly resilient, and a survey of an additional 300 people.[8] Their research identified five key factors that the most resilient people possess, namely:

- **optimism** – that is, having a clear vision of an inspiring and realistic future, and the self-esteem and confidence to believe you can realise it;

- **freedom from undue stress and anxiety** – that is, knowing where your own tipping point is and having effective strategies to both distract from and resolve the source of stress;

- **taking personal responsibility** – that is, learning to spot and manage your 'drag anchors' or limiting beliefs, being able to reflect and learn, knowing how to deal with conflict effectively and when and how to ask for help;

- **openness and adaptability** – that is, a willingness to change and an ability to do so while remaining authentic and true to yourself; and

- **a positive and active approach to problem solving** – that is, developing and trusting both your rational and your intuitive approach to decision making, and knowing when and how to seek input from other.

Other research has indicated that **compassion** is also a signature characteristic of resilient people and that cultivating compassion is a key factor in building resilience.[9]

All of the above characteristics can be learned and developed, and organisations can support people in this in order to build resilience. Much of what has already been discussed elsewhere in this book will help in this context – again comes the circular idea that resilience and energy are both a product of a well-managed change process and a necessary prerequisite for one. Telling a compelling story will help to build optimism per se by helping people to conceive a clear vision of the future – and see below for the many other ways in which stories support resilience and energy. Creating a strong sense of belonging and building confidence will also contribute to self-esteem and self-belief and, by extension, to optimism.

Taking an evolutionary approach to strategy setting and prioritising, creating the conditions for agility and innovation, and actively prioritising the deepening of understanding, inclusion and the development of networks will all support people to take personal responsibility, remain flexible and open in their outlook and proactively solve problems.

Organisations also have an important part to play in creating a highly flexible context within which people can balance their working lives and the rest-of-life, and in helping people to manage any stress or anxiety. This can be a tough nut to crack, particularly in a professional services organisation full of achievement-oriented people who will tend to go the extra mile, give a lot of discretionary effort, and have an innate tendency towards perfectionism. That said, it is vital that organisations have as much regard to their peoples' mental health as to their physical health. At CMS, as the new firm established its identity, it took the opportunity to enhance its existing approach in this area by developing what was a ground-breaking approach to wellbeing, that it can rightly be very proud of. The new firm put in place policies which are market-leading in their support for family life and flexibility. It also sought to actively support and promote mental health. For example, the new firm engaged a psychotherapist to be available to the whole firm. The firm also worked with its main healthcare provider to develop an end-to-end mental health protocol which was just as comprehensive and detailed as its physical health protocol, and developed a battery of online resources and learning programmes for people to use. A cohort of 'mental health ambassadors' were trained in how to spot signs of stress and offer support. The pilot of this training was oversubscribed by four times the number of places, and so the firm ran additional programmes and now has teams of mental health ambassadors across every part of its UK business. The firm also coached the whole leadership team in the same skills around how to identify and address potential issues and how to access support for people in their teams. Here is a perfect example of purpose being lived out, having regard to their mental health being an integral part of CMS's purpose in 'creating sustainable and rewarding futures for our people'.

Building Energy

People and organisations who can truly master the various aspects above – optimism, responsibility, problem solving, openness and adaptability – and who have learned how to manage their stress levels may very well find themselves to be more than resilient; they may find themselves to be actively energised by the role and context within which they find themselves.

But what would it look like to go further into this idea and find the means to enable people and organisations to actively thrive precisely by virtue of their participating in change?

There is evidence that participation in a community in pursuit of a common goal is actively energising – this is the 'we're all in it together' argument. Much of Alex Haslam's work on social identity[10] is based on his central research finding that we define ourselves primarily in relation to other people; our sense of shared identity is central to our entire sense of identity. Once we identify with a particular social group, says Haslam, we will proactively work with the group – that is, we will expend discretionary effort – in order to achieve group outcomes. A study involving older adults living in a care home[11] found that those in a group who were consulted collectively about the décor of the communal areas were found to identify more with staff and fellow residents, display enhanced citizenship, and have improved cognitive skills, lower incidence of depression and better physical health than a control group who were given a newly decorated lounge but were not consulted. This suggests that, done well, participation in community with others in a change project can be actively beneficial for people.

A second aspect of how change can be actively energising is the 'momentum argument'. This echoes Mark Humphries' observation earlier that some people get on the bus just because they like the idea of moving. Certainly in my experience, and within reason, successful change can act as a catalyst for more change – partly by building on the confidence that it can be done, partly because some change can ignite the ambition for more, partly because once people are in the frame of mind to embrace change it makes sense to do what needs to be done and, yes, partly, because some people do simply enjoy the ride.

One example from the CMS merger experience relates to the Front of House team – reception, security, audio visual support and catering. Prior to the merger, this team operated well. It was full of great people, most of whom who had been with the legacy CMS firm for a while and knew everyone in the business. The systems supporting the team were not market-leading, but they were fit for purpose, and good relationships kept the whole thing on the road. The merger challenged this equilibrium. We saw unprecedented demand for the use of our meeting rooms, and the team could no longer rely on personal relationships because they did not know everybody any more. At first, the response was simply to patch things up, but it quickly became apparent that there was a clear ambition – a real fire in the belly – on the part of all concerned to make the Front of House experience as good for clients as it could possibly be. The firm engaged coaches and hospitality experts to support the team, sought feedback and looked at best in class experiences in other sectors. The team changed the physical layout of the reception area, invested in new systems, changed the team structure, and engaged everyone in the firm in developing a 'playbook' or protocol that everyone agreed to follow. The firm did not 'need' to do this. It was never in the plan, and it cost time and money. But it was absolutely the right thing to do. The merger drove change and enabled this further change programme that further improved the firm and its ability to delight clients.

A third way to think about change and how it can be an energiser is to think about change as a creative process. This lends a fresh perspective. In *Making Ideas Happen*, Scott Belsky focuses on an artist called Jonathan Harris in exploring how creative people and entities shift from ideas to execution.[12] At one point Belsky and Harris talk at length about love, and about the paradoxical role it plays in any creative pursuit. At first, love drives us forward – we are interested, we want to focus and learn and improve. When things get tough, love keeps us going. That's resilience. But love can also disappoint us. Harris says:

> . . . The thing you actually end up making is going to be such a failure compared to the original vision that you had . . . when you really fall in love with something, you idealise it and you develop a vision of

it that's actually unattainable in reality. The feeling of it is so pure that you can't make a real thing that has that feeling, and so you're inevitably going to be disappointed by it . . .

I chewed on this quote for a long time. Love is a strong word in a business context, and failure isn't wildly popular as a concept either. But I think what Harris is getting at here gives us a critical insight into what it takes to thrive on change: namely, that we ultimately have to be in love with the creative process itself, as much as we are in love with any particular vision for the outcome.

My five-year-old son loves to draw. Our fridge is covered with elaborate dinosaurs, giraffes, ant colonies, maps. But just recently, in the past few weeks, he will sometimes finish a drawing and, rather than brandish it proudly, will burst into floods of angry tears and scrunch it up because 'it doesn't look like the thing at all'. He has hit some sort of developmental milestone that enables him to perceive the gap between his vision for the drawing and the reality, and it frustrates him.

Clearly, in art and in business, there does need to be some kind of vision for the outcome. The leadership team knew at the outset of the CMS merger what they wanted to create – indeed, the vision for the new firm was a critical component in building consensus and in inspiring and driving the change process. But we have already talked at length about the need to take an evolutionary approach rather than being didactic, and about the commercial imperative to remain agile. We have also talked about the need for collaboration and collective leadership, and the importance of allowing space for people to develop their own stories and their own expression of the corporate purpose. All of these factors militate against an approach which would conceive every detail of a theoretically perfect firm and then focus entirely on delivering it. No two people would ever conceive of exactly the same firm in any event – remember the twenty different still life drawings of a jug of orange juice. The writer Elizabeth Gilbert, talking about the creative process, says:

> ... the most interesting part of the entire engagement is not necessarily what you end up making. It's what making that thing does to you.[13]

The Role of Purpose

What role does purpose have to play here? How can a clear sense of individual and organisational purpose help us to build resilience and energy – perhaps even to find change actively energy-giving rather than depleting? Friedrich Nietzsche famously said:

> He who has a why to live can bear almost any how.

More than fifty years later, we now have research to support the argument that a clear sense of purpose can increase resilience.[14]

At an organisational level, a recent research report by Tomorrow's Company in collaboration with Danone, which looked at the value of a purpose-driven approach across twenty global companies found that one of the key ways in which purpose generates value is by becoming a source of energy and commitment for people within the organisation.[15] Jason Barnwell, Assistant General Counsel for Legal Business, Operations and Strategy at Microsoft, whose crystal clear articulation of his personal purpose was quoted in Part One, testifies to this. He says:

> a clear sense of purpose activates the creativity and discretionary effort in my team – and that's where excellence comes from. I see more autonomy, more creativity. I see people stepping up, stepping out, doing more.

Jason then talks about how he ensures that he and his team consistently deliver what he calls 'super-normal' performance by having constant regard to three things – purpose, capabilities and opportunities, and ensuring that all three are kept aligned and in balance.

It is tempting – but a mistake – in thinking about an energised organisation to imagine a happy lighthearted place; often, the most energised organisations are focused on solving hard problems together. Think about when you last felt most deeply engaged in a work context. How did it feel? Here is Anne Brafford, a former US lawyer, talking about preparing for her first appellate argument:[16]

> During the many days of preparation leading up to my court date and during the argument itself, I felt excited, fully absorbed, invested in working for long periods, and challenged but ready to be resilient. The case was important to my client and to me. I was fully engaged.

This certainly chimes with my own experience. Those days which have been amongst the hardest days of my professional career have also been the most rewarding, partly because the cognitive challenge of work is itself a key part of job satisfaction and work engagement,[17] but mostly because I was utterly committed to the higher purpose both of the work itself and the outcomes we were working towards.

These experiences – yours, Brafford's and mine – give the lie to what Quinn and Thakor call 'the largest barrier to [organisations] embracing purpose' [18] – namely, the cynical transactional view of employee motivation that says we are only interested in minimising effort and maximising (financial) reward. Paul Dolan in *Happiness by Design*[19] argues that purpose is in partnership with pleasure as two components of happiness. If people have a clear sense that their work is purposeful, they will to some extent trade a degree of pleasure in pursuit of the purpose. This is a better framing than 'delayed gratification', which implies that there's nothing in this for me, now. If I am feeling purposeful and engaged, that in itself is the 'something in it for me'.

The Role of Stories

In a sweet little vignette reminiscent of the apocryphal 'I'm building a cathedral' story, I recently took my four children to the dentist. Here are their reasons for going, as expressed by them when I asked them afterwards, 'Why did we go to the dentist this afternoon?':

'Because we don't have gymnastics or piano lessons on a Tuesday.'

'Because we get stickers.'

'Because it is important to look after our teeth and the dentist can make sure they are healthy.'

'Because I like the dinosaur books [in the waiting room].'

Same trip, same dentist, four different reasons for going, and so – with a bit more digging – four very different stories about what we did that afternoon as a result.

When the children are a little older, and responsible for taking themselves to the dentist of a Tuesday afternoon, the research says that those of them with a narrative around taking care of their teeth which has a higher level of construal – where the why, the purpose, is at the fore, rather than the how, and where the brain is actively using facts it already knows and using these to influence its action – those are the children most likely to stick with going to the dentist![20] One key role of stories in keeping us resilient and energising us is, once again, that they keep us connected to our purpose.

Another role of stories in this context is that they help us to imagine a route through. By recalling times in our own lives where we have faced similar challenges, or by hearing others' stories, we can remember successes (and increase our dopamine), build our confidence, stand on our cairn (or someone else's), regain a sense of control and autonomy (and reduce cortisol), and be optimistic about the future. Homer already knew this:

Be strong, saith my heart. I am a soldier, I have seen worse than this.

Notes

1. Dinesen, Isak (pen name of Karen Blixen), in an interview with Bent Mohn in the New York Times Book Review, November 1957.
2. Whyte, David, *Crossing the Unknown Sea: Work As A Pilgrimage of Identity*, Riverhead Books, 2002.
3. Rock, David. *Your Brain at Work*, HarperBusiness, HarperCollins, 2009.
4. Cohen, S., Janicki-Deverts, D., and Miller, G.E., 'Psychological Stress And Disease', *JAMA*, 298(14): 1685–1687, 2007.
5. Scarlett, Hilary, *Neuroscience for Organizational Change: An Evidence-Based Practical Guide to Managing Change*, Kogan Page, 2016.
6. In his commencement address before the 1996 graduating class at the University of North Carolina at Chapel Hill.
7. Scarlett, Hilary, *Neuroscience for Organizational Change: An Evidence-Based Practical Guide to Managing Change*, Kogan Page, 2016.
8. Clarke, Jane and Nicholson, Dr. John, *Resilience: Bounce Back From Whatever Life Throws At You*, Crimson Publishing, 2010.
9. Fernandez, R., '5 Ways To Boost Your Resilience At Work' *Harvard Business Review*, June 2016.
10. For example his lecture on Social identity and the new psychology of mental health, at: https://www.youtube.com/watch?v=TWWZd8lrraw, watched 1 July 2018.
11. Knight, C., Haslam, S.A., Haslam, C., 'In Home Or At Home? How Collective Decision Making In A New Care Facility Enhances Social Interaction And Wellbeing Amongst Older Adults' *Ageing and Society*, Volume 30, Issue 8 November 2010 , pp. 1393–1418.
12. Belsky, Scott, *Making Ideas Happen: Overcoming The Obstacles Between Vision And Reality*, Penguin, 2011.
13. Elizabeth Gilbert, in discussion with Reid Armbruster, for the blog at www.audible.com about her book *Big Magic*, 10 August, 2016, read on 12 August 2018.

14. Rutten, B.P., et al (2003) Resilience in Mental Health: Linking psychological and neurobiological perspectives. *Acta Psychiatrica Scandinavica* 128(1): 3–20.
15. Tomorrow's Company Report, with Danone: *The Courage Of Their Convictions: How Purposeful Companies Can Prosper In An Uncertain World*, Mark Goyder and Norman Pickavance, 2018.
16. Brafford, Anne, *Positive Professionals: Creating High-Performing Profitable Firms Through The Science Of Engagement*, ABA Law Practice Division, 2017.
17. Crawford, E.R., Rich, B.L., Buckman, B., Bergeron, J., 'The Antecedents And Drivers Of Employee Engagement' in Truss, C., Delbridge, L., Alfes, K., Shantz, A. and Sloane, E. (eds.) *Employee Engagement In Theory And Practice*, Routledge, 2014.
18. Quinn, Robert E., and Thakor, Anjan V., Creating a Purpose Driven Organisation: How To Get Employees To Bring Their Smarts And Energy To Work, *Harvard Business Review*, July/August 2018, pp78–85.
19. Dolan, Paul, *Happiness by Design*, Penguin 2014.
20. Fujita, K., et al, 'Construal Levels and Self Control', *Journal of Personality and Social Psychology*, 90 (2006), 351–367 and Cable, Daniel, M., *Alive at Work: The Neuroscience of Helping Your People Love What They Do*, Harvard Business Review Press, 2018.

Part Four

Implications

Chapter 16

The Bigger Picture

When we try to pick out anything by itself, we find it
hitched to everything else in the universe.

John Muir[1]

In the course of this book I have shared my recent personal experience of organisational change, and that of others. By examining those experiences, with the benefit of hindsight and in the context of the available literature and research, I hope I have shown how an approach to change which is rooted in an organisation's purpose and depends on storytelling as a key part of its methodology can help an organisation to tackle the various challenges which periods of change tend to present – and even to thrive by virtue of the process of change itself. I hope I have shown how purpose and stories can help to engender a sense of belonging and identity; how they can enable organisations to take an evolutionary approach to strategy setting and implementation; how they build confidence, and foster agility; how they create understanding, allow simplicity and preserve and build energy.

The vital question now, having established their usefulness in achieving the BECAUSE in a change context, is whether these concepts of purpose and storytelling bear up in a wider business context. This is critically important because leaders need to lead in a way that has regard to the cohesion and congruence of their whole organisation, both within itself and in relation to its stakeholders and the market. The BECAUSE theory is not a discrete tool that can be applied like a topical poultice to a change project without any ramifications for the wider organisation – it is a set of principles, an underlying philosophy that, taken like intravenous medicine, will permeate the entire system. On that basis, leaders should only prescribe it if they believe it to be of overall benefit to their organisation. In the starkest commercial terms, if by pursuing a purpose-led approach you are pursuing something that the wider market does not want, and so will not pay for, you are diluting value.

In seeking to answer this question, we should consider two things – first, what we know or can infer about the wider applicability and implications of the BECAUSE approach itself; and, secondly, what we know about how stakeholders define good performance, what they want, and how the BECAUSE approach may contribute to this. The first limb, I can, and will, speak to with some authority and a lot of passion; the second limb I touch on comparatively briefly, offering just a few observations and ideas, largely because to comment with any authority would require a detailed understanding of the particular industries and markets in which organisations are doing business.

BECAUSE At Large

The first limb is fairly easy to address by briefly revisiting some of the observations made earlier in this book and considering them through a wider lens, noting first of all that we have alluded throughout to the context within which pretty much every organisation exists, which in a sense makes change a constant. And to be clear, I am not thinking here exclusively, or even primarily, about profit-making businesses – volatility, uncertainty, complexity and ambiguity impact on government, public sector organisations, schools, charities and other not-for-profits, even social groupings – societies, clubs, faith groups, families. Sam Baker, a partner in Monitor Deloitte, Deloitte's strategy consulting practice, and a leading thinker in the field of purpose, says that the real litmus test for the value of a purpose-led approach is whether it delivers value day-to-day in the everyday processes and transactions of an organisation. Although we have focused in this book on specific examples of change in relation to almost every aspect of the BECAUSE theory, we have also touched on its relevance in the context of the wider VUCA factors, which are present every day and form the backdrop against which every organisation operates.

There is a wider point still. As I said at the outset, the central thesis of this book is quietly revolutionary. If BECAUSE works, and I believe it does, then it is a strong advocate for the case for a whole approach to doing business which is about vulnerability rather than power; about inspiration rather than compliance; about engagement rather than control.

In terms of considering more specifically the relevance and importance of BECAUSE across the wider organisational context, let's take each of the seven concepts in turn. Consider **Belonging** first. A sense of belonging always matters – that chapter of the book opens by presenting the compelling evidence around how a sense of belonging is the single biggest predictor of health and longevity. The chapter then goes on to demonstrate how a purpose-led approach, and a really committed and vivid approach to the creation and telling of stories, can help to bond and align leadership teams, create connectedness amongst teams on the ground and across remote teams globally, create 'tribes' and a sense of identity and build a sense of physical and psychological safety. All of this is vitally important all the time, and not just during a period of change.

Next up, **Evolution**. This chapter explains that during periods of profound change, an evolutionary approach to strategy setting is vital – the degree of uncertainty and complexity is such that an overly-planned and didactic approach in these circumstances would be just plain daft. But this chapter also argues that an evolutionary approach is actively desirable because it enables swift decision making, allows flexibility in the face of complexity or the unexpected, empowers people and builds engagement, and enables better-quality 'coalface' decision making to take place – all nice-to-haves for a business in any context. The chapter explains how a purpose-led approach, enriched by storytelling, lays the ground work for this sort of evolutionary approach by enabling leaders to be vulnerable, fostering hope, encouraging reflection and learning and igniting curiosity.

The case made for building **Confidence** during times of change is that it enables organisations and people to 'transcend and include', to embrace new ideas, to make swift decisions *con brio*, to hold their nerve and to manage conflict and to play. Again, none of these strike me as being relevant only during periods of change. The approaches identified which help to build Confidence are to reframe challenges, to 'build your cairn' of past achievements and to stand on it, to stay rooted in what you hold most dear, and to take your space – physically, and in time. All of these approaches call for stories, and all are rooted in purpose.

The argument in relation to **Agility** is a slightly different one – this chapter explicitly draws in the wider societal and market context and argues that, while they are changing anyway, organisations need to design for agility because it is fast becoming a prerequisite for all organisations in today's environment. The chapter uses the metaphor of a gymnast – with a strong backbone, yet incredibly bendy everywhere else – and shows how purpose and stories both form that strong backbone and enable the flexibility and dynamism required across the business. Purpose is the 'north star', and purpose and stories together enable dense networks of aligned and accountable people to bond, work together, make decisions quickly and learn fast. Purpose also underpins a dynamic and empowering people model and helps to ensure that technology is used effectively as an enabler.

The chapter on **Understanding** talks about the skills and mind-set required to engender deep understanding as between people in an organisation and as between an organisation and its stakeholders – again, something which is important at all times and not just during periods of change. It explains how a clear sense of purpose can support real clarity of communication at every level, and also how, along with a rich bank of stories, it can foster much deeper understanding beyond clarity – what I call 'enlightenment' – building trust, embracing difference, establishing common ground and supporting high performance.

The chapter on **Simplicity** opens with a lament about how we are all increasingly drowning in complexity. This condition may be exacerbated during periods of change, but it is ever present. The chapter offers an approach rooted in purpose and stories which will help organisations to achieve greater simplicity of message, of information, of decision making and execution, of structure and rules and of expectations; an approach that can be applied in the face of complexity at any time.

Finally, when it comes to building resilience and **Energy**, this book shows how an approach rooted in purpose and supported by stories which have the highest possible level of construal can build optimism, help people to manage stress, build responsibility, increase openness and positivity in decision making, energise people through a common goal, build momentum and foster creativity. Once again, while energy and resilience undoubtedly face particular challenges during periods of profound change, every organisation and every leader is interested in maintaining resilience and maximising energy at all times.

It is clear, then, that the approach that this book advocates around being purpose-led and story-driven is capable of delivering benefits in relation to each of the elements of BECAUSE more generally for organisations, and not just during periods of change.

Stakeholder Perspectives

The second consideration, then, is whether this purpose-led, story-driven approach is something stakeholders are asking for, and whether it is value-generative? Every organisation will have to answer this question for itself, in relation to its own stakeholder group in its own market context – it requires a sophisticated reading of the way your own particular world is moving. I will offer just a few observations based on my own experience, my research in the course of this book and my work with Business In The Community on corporate purpose in general and on the UN's Sustainable Development Goals (SDGs) in particular. Let's touch on the perspective of four key stakeholder groups – talent, customers, regulators and government, and investors.

In relation to talent, first, in my opinion the case for a purpose-led, story-driven approach is squarely made. Throughout this book there are myriad examples and research findings to support the argument that this approach builds engagement and motivation, energises people and builds their resilience, develops psychological safety, drives high performance and fosters creativity. A proliferation of studies and surveys show that the latest generations to join the workplace are acutely aware of this, and actively base decisions to join, remain and leave organisations on their assessment of how purpose-driven an organisation is. For example, a 2014 study in the US across 300 organisations found that 94% of millennials want to use their skills to benefit a good cause.[2]

In relation to the customer group, I consider the case to be less clear, certainly in the professional services sector that I know best. I have no doubt at all that many of the consequences and benefits of a purpose-led approach support organisations in delivering good service to their customers; we know that clients want to work with strong teams who are agile and energised and empowered to solve their problems, who listen well and communicate with real clarity and understanding, and who are not ensnared by process.

But 'purpose' for many is still backstage in their interactions with their advisors – most are not explicitly asking us about it, or placing a value on purpose or 'pricing it in'. My hunch is that this will change. It is now entirely expected that firms will share details around their diversity and inclusion policies and statistics and sometimes work with clients on joint initiatives in this area. The same is true in relation to Corporate Responsibility commitments and initiatives. I think that organisations will, in the short to medium term, start having equivalent discussions with their advisors about purpose. As organisations grapple with purpose for their own organisations – many are working, for example, to embed the SDGs in their strategy – they will be keen, I think, to extend this new orientation and thinking down their value chain to their advisors and suppliers and to ensure that they are aligned in this area.

In business-to-consumer businesses, there is undoubtedly a trend towards increased awareness of, and demand for, organisations to be purpose-led. Keith Weed, Chief Marketing Officer of Unilever, a company that is leading the way in this area in its sector, says:[3]

> Consumers are increasingly looking for – and expecting to see – the purpose behind the brand. The expectation not just that they won't 'be bad' but that they should actively 'do good' is not going to go away.

That said, my sense is that, nevertheless, purpose-led purchasing decisions remain the preserve of the privileged few because of the affordability – or perceived affordability – of the brands that currently go to market with their purpose-led agenda explicitly out there. Purpose currently seems to be associated with premium pricing – perhaps because it is enhancing brand positioning, or perhaps in order to cover increased costs of production associated with being purpose-led. Many organisations speak to purpose, but relatively few seem to be as advanced as Unilever in their capacity to deliver. In a knowledge business like a law firm, it is easy to assume that there are minimal, increased costs associated with a purpose-led approach (increased investment in leadership development, for example), and that these are more than offset by the increases in productivity, efficiency and quality of decision making that will result. In a different context – mass-production of clothing, for example, or farming, or pharmaceuticals – we can absolutely see that the equations fall differently and require choices to be made, at least in the short-to-medium term as organisations adjust their operating model to accommodate a purpose-led approach. Organisations will likely only make – and arguably *should* only make – these adjustments if they are persuaded that this approach will ultimately be value-generative across all stakeholder groups. Meanwhile, organisations should be wary of passing on any increased costs of production to consumers before their understanding of, and demand for, a purpose-led approach is sufficiently developed: this would exacerbate the risk that purpose comes to be (mis)understood primarily as a marketing tool, a strapline, a brand-enhancement that few can afford to pay for, and which can be readily abandoned in austerity, and the value of purpose in the consumer relationship will be rejected before it has a chance to develop and be proven and understood.

The position in relation to governments and regulators as stakeholders is also complex. Philosophically, certainly in the West, the post-2008 zeitgeist is all about a more ethical and transparent approach to doing business; a worldview that sits well with a purpose-led approach. There are many examples of government driving the agenda in this area: for example, European, Asian and African governments have now all issued sovereign green bonds and the French government implemented the first ever mandatory environmental impact disclosure rules for asset owners in 2016.

Globally, the establishment of the UN's seventeen SDGs, with a target date of 2030, has significantly increased the noise around responsible business and responsible investment as the drivers of business growth, and has given businesses globally a common agenda and a common language. The work that many of the biggest global companies – Unilever, Coca-Cola, Deloitte – are now doing as a result to embed the SDGs into their strategy means that they are very explicitly putting purpose right at the heart of their core business model. In doing so, they are also strengthening the enabling environment for doing business and building markets around the world. All of this means that purpose should be viewed favourably by governments and regulators, and it generally is.

I do perceive a tension, though, both around timeframes and around means. National governments driving a short-term growth agenda in the context of austerity may still be primarily concerned with increasing productivity over a shorter timeframe than the purpose agenda contemplates. As in relation to consumers and investors (see below), the issue here is around balancing now and later; today and tomorrow. Businesses need to ensure that they are driving the agenda, without getting too far ahead of the curve.

In one very clear and illustrative example of the increasing emphasis that national regulators are placing on purpose and sustainability,
the new Corporate Governance Code which took effect in the UK on 1
January 2019 places a new and very clear emphasis on a company's role
in society and stresses the need to engage with all stakeholders, including employees and wider society. The Code explicitly states that it is the
role of the board to determine the purpose of a company and ensure the
company's values, strategy and business are aligned to it.

More generally, in terms of means and approach, regulators tend
to drive the behavioural changes and accountability they require via
increased regulation (of course!), supported by rigorous reporting require-
ments and sometimes elaborate processes and systems, all of which militate somewhat against the trust-based, process-light, agile approach that
I have advocated for throughout this book. In the vast majority of sectors,
I do not, however, believe this to be a deal-breaker for the purpose-led
approach; it simply requires it to operate within a particular pre-existing
framework around compliance.

The case in relation to investors takes us right back to Part One of the
book and the various definitions of purpose and its relationship with
value creation. While writing this book I have pondered whether it is
somehow easier to adopt a purpose-led approach within a partnership
structure, such as most law firms have, or perhaps within a shared ownership structure, such as Arup, for example, has, rather than within a
context which has external investors. On balance, I have concluded that
this is not the case. There might be some advantage in terms of the leadership's capacity to demonstrate the value of a purpose-led approach
to having your investors essentially inside the business, experiencing
the approach first hand. But exactly the same investor concerns around
transparency, accountability and, ultimately, return on investment apply,
as does the same dynamic tension between maximising immediate
return and creating long-term sustainability, arguably all the more so
in a law firm which is used to distributing all its profits every year. And

the environment is such that they are Right There, literally standing over your desk in open plan, holding the leadership to account. Arguably, a partnership environment is the ultimate crucible in which to prove the case for a purpose-led approach.

That said, it now seems clear that investors of all kinds are increasingly interested in the wider consequences of their investments, and are coming to embrace a more holistic and long-term definition of value creation that incorporates – indeed, depends upon – a purpose-led approach. Larry Fink, Chairman of Blackrock, one of the world's largest investment management firms, titled his 2018 annual letter to CEOs 'A Sense of Purpose',[4] and stated:

> Society is demanding that companies, both public and private, serve a social purpose . . . Without a sense of purpose, no company can achieve its full potential. It will ultimately lose the license to operate from key stakeholders. It will succumb to short-term pressures to distribute earnings and, in the process, sacrifice investments in employee development, innovation, and capital expenditures that are necessary for long-term growth . . . ultimately, that company will provide subpar returns to the investors who depend on it . . .

In an insightful blog article for the Hit and Run blog on reason.com,[5] happily published on the very day I was working on this chapter (everyone gets lucky sometimes!), Nick Gillespie revisits the controversy that surrounded the publication of Mackey and Sisodia's *Conscious Capitalism* in 2013,[6] and reflects on how far the debate has come in five short years. Mackey's statement in a 2013 interview[7] that:

> Our book challenges the notion that stakeholders in business are inherently opposed to one another. That is not the case. When businesses operate with higher purpose beyond profits and create value for all stakeholders, trade-offs are largely eliminated, performance is elevated and the entire system flourishes. Everyone wins.

no longer sounds whacked-out hippy or even terribly controversial, and we are increasingly seeing investors putting their money where Mackey's mouth is.

Notes

1. Muir, John, *My First Summer in the Sierra*, Houghton Mifflin, Boston, 1911.
2. Achieve Consulting Inc., *Millennial Impact Report*, June 2014.
3. Business In The Community, *Inspiring Businesses To Improve Society Through Purpose-Driven Brands: A Purpose Toolkit*, January 2018.
4. Fink, Larry, *Annual Letter to Shareholders 2018: A Sense Of Purpose*, published March 2018 and available at https://www.blackrock.com/corporate/investor-relations/larry-fink-ceo-letter.
5. Gillespie, Nick, 'John Mackey and Conscious Capitalism Have Won the Battle of Ideas With Everyone but Libertarians', published on the Hit and Run blog on Reason.com on 15 August 2018 , accessed on 15 August 2018 (!)- https://reason.com/blog/2018/08/15/john-mackey-and-conscious-capitalism-hav.
6. Mackey, John and Sisodia, Raj, *Conscious Capitalism: Liberating The Heroic Spirit Of Business*, Harvard Business Review Press, 2014.
7. Schawbel, Dan, 'John Mackey: Why Companies Should Embrace Conscious Capitalism', interview with John Mackey published on Forbes.com on 15 January 2013, accessed in June 2018 at https://www.forbes.com/sites/danschawbel/2013/01/15/john-mackey-why-companies-should-embrace-conscious-capitalism/.

Chapter 17

The Best Thing You Can Bring Is Heart

Here is the world. Beautiful and terrible things
will happen. Don't be afraid.

Frederick Buechner[1]

In closing, it seems important to consider, in the light of all we have learned, the key takeaways in terms of how we should aspire to lead, and how we can encourage and equip future leaders of purpose-led, story-driven organisations so that they, their organisations and their people can all thrive.

In 2014, in partnership with Cranfield University's Doughty Centre for Corporate Responsibility and *The Financial Times*, Coca-Cola commissioned research into current and future business leaders' opinions about the relationship between profit and purpose – now and in the future.[2] The results are striking. All current and future leaders (people currently operating in the level immediately below the C-suite) who participated in the survey agreed that a purpose-led approach builds engagement, increases trust, fosters innovation and increases an organisation's relevance to future generations. Yet while nine out of ten current CEOs believe that their organisation has a clear purpose, only two out of ten future leaders agree, and 'management attitudes' is identified as one of the two biggest barriers to the successful adoption of a purpose-led approach (the other being government policy). Clearly, then, there is a gap between intention and perception that leaders will need to bridge if they are to truly embrace purpose. The stakes are high. Larry Fink again in his 2018 letter to CEOs:[3]

> Today, our clients – who are your company's owners – are asking you to demonstrate the leadership and clarity that will drive not only their own investment returns, but also the prosperity and security of their fellow citizens.

That there is sometimes a gap between intention and perception, ambition and execution, should come as no surprise – the demands on leaders today are huge, varied and complex, and coming at them at speed, and leaders may be ill-equipped to navigate their way through and set a clear path for others. This second point is perhaps particularly true in the context of professional services organisations, which have historically promoted on the basis of technical legal expertise and market standing and have chronically underinvested in developing leadership skills in their people.[4]

But what is required here is, I think, more than some patching up to close the gap. We need a new paradigm – an open, vulnerable, questing and questioning, creative form of leadership. This is a high vision, a big ask, and not a little daunting.

Not Another List

Now. I have read enough management books in my time to know that this is the point at which a book like this can get a little dispiriting. There will follow a long list of the things that leaders *should* do, all of which I will heartily agree with and underline, and fully intend to implement. . .I might even fold down the corner of the page so that I can find the list again easily. . .and then real life comes bursting back in, and all of the pearls of wisdom I've underlined become just Yet One More Thing on my never-ending to-do list.

Instead, then, I am going to conclude this book by highlighting just three broad perspectives or areas that leaders may wish to focus on – without distilling anything down into lists, so you can put that pen away. Then, in the true spirit of this book, I am going to end by telling you a story about just one leader and how she makes purpose live in her organisation.

Three Thoughts

Leaders, then, should consider their role and contribution across three broad areas – the nature of the work that their organisation does, the people within it, and the environment within which those people operate. In all three of these areas, leaders can have an impact which will help their organisations and their people to become more purpose-led and better able to thrive through BECAUSE and beyond.

First, then, **the nature of the work**. As we explored in Part One, leaders have a key role to play in helping organisations to discover their purpose and then in telling the central stories that keep that purpose alive. Leaders light the fire. Many people talk about burning platforms, others talk about burning ambition.[5] I like the latter much more as it draws people forward, instead of simply inciting fear and panic. It makes me think of the pillar of fire, the manifestation of the presence of God, which guided the people of Israel by night on their journey to the promised land; a leader's presence bringing guidance, safety, light and warmth. Or remember the campfires around which our ancestors gathered to tell stories? Leaders light the fire around which the organisation can gather and tell its stories. Leaders can help people to see that their work matters by placing it in the context of the big story, highlighting its social impact, marking and celebrating progress and ensuring that everyone remains challenged and stretched.

Secondly, having established that the work matters, leaders should consider how best to show their **people** that they matter too. They can do this by giving them autonomy and independence, trusting them, involving them, and enabling them to achieve. Leaders need to have a growth mindset, and to encourage the same in everyone in their team. Jason Barnwell, in talking about how he helps to foster a sense of purpose in his own team, spoke to me about how he, too, is influenced and inspired directly by the leaders above him in Microsoft. He talked about Satya Nadella, Microsoft's CEO, who very explicitly focuses on maintaining his own growth mindset and inculcating the same across the whole global business. Nadella has written about how he has focused on creating an inclusive environment where everyone can bring their best self to work,[6] and Barnwell testifies to having experienced this personally, for example on a recent conference call to run through a presentation during which

Nadella specifically commented on the quality of the presentation documents, acknowledged the work involved in preparing them and publicly thanked those on Barnwell's team who had done that work.

Recent research[7] shows that the leaders who are best able to engage their teams and organisations in the organisation's purpose share four personality traits in particular:

- they are curious and inquisitive – asking questions and looking for new ideas in a way that engages people in thinking and discussing;

- they are challenging and relentless – but not erratically, and not too much – the best leaders are restless and need to be moving, and make people want to move with them;

- they hire for values and culture fit; and

- they are able to trust people.

Finally, leaders can think about the **environment** within which people are operating and consider how to rig it to support more focus on purpose and to increase engagement. Robert Snyder urges, 'before you go to P [i.e. people], think about the E',[8] and says that it is generally easier to change the environment than it is to change people. Subject to the small matter of VUCA, I agree with this – leaders can think about some of the issues we addressed in the chapter on Simplicity, and consider which of their rules, systems and processes truly support their business and which get in the way. They can think about what they measure and reward, what they celebrate, what stories they tell. They can think about how the physical environment is arranged to support belonging, how they structure teams, and how they cascade information throughout the organisation. . .

But before this too risks becoming a list, let me tell you the story of Jandel Allen-Davis, as just one example of leading for purpose.

Jandel Allen-Davis

Jandel Allen-Davis has been a physician for 34 years, specialising in obstetrics and gynaecology. She has a manner that embodies a combination of deep-flowing compassion, expansiveness of time and possibility, and crystal clarity of thought. Talking with her feels like sitting by a river – I immediately relax in her presence.

In 2006, Jandel was on the senior team of the Colorado Permanente Medical Group. In that role, she was partnered with the Kaiser Foundation Health Plan leadership team, looking at the myriad causal connections between health plan provision and the frontline delivery of medical care.

Kaiser Permanente is the leading not-for-profit health plan in the US, serving more than 11 million members across 600 locations in nine states. It has its roots in the Great Depression – one surgeon and a twelve-bed hospital in the Mojave Desert, in the shadow of the Colorado River Aqueduct Project. It has been providing high-quality, affordable healthcare to its communities ever since.

Jandel herself is based in Colorado, and in 2009, she became VP for Government, External Relations and Research at Kaiser Permanente – an unusual move in a world where doctors rarely leave medical groups to join the health plan side. And a huge personal move for a physician who loved patient care and had never before had a leadership role in a corporate environment.

Jandel's team had responsibility for communications and brand, direct community benefit investment and local government and stakeholder relations. Together with the sales and marketing department, they promote the value of an integrated delivery model as one important answer to the changes which healthcare providers and communities in the US are currently grappling with in a post-Obamacare context.

Jandel says:

I've always loved the saying 'It's a lot easier to repair the roof when the sun is shining'. When I arrived at Kaiser Permanente, it's fair to say that there were roof repairs required! But you don't just sweep in there like a tornado and start ripping things off.

I had to take a step back and think, 'What do I bring?' I'm a doctor. I am good at getting to know people. So I took the time. I asked them Why? – why do they do what they do? I asked them What if. . .? I asked them about their dreams for this work. What were their ideas for our patients, our members, our communities?

I found that when I did it that way, I got high, high levels of engagement. People were willing to go the extra mile, to take risks. I was helping them to develop their agility – they were being stressed and stretched appropriately so that when the tough times come – and come they always do – we're ready for them.

I've learned that as a leader, you've got to exercise the muscles of compassion and listening constantly. You've got to be helping people tap into their purpose and crafting stories with them. These particular dimensions of strong leadership have relevance and need to be leveraged all the time, so that when change happens you're ready for it. It's like training for a marathon.

So, even when challenges we've seen before come around again and again we say, 'Okay, what do we already know? What can we do? Let's just saddle up and get it done. Let's face the music joyfully.' Change resilience is sorely lacking in this sector.

I asked Jandel what, specifically, she does as a leader to really foster this kind of resilience and positivity in her team. She tells me about a conversation with a former colleague that takes place – like so many of the best conversations – in the ladies' loo.

'What's cool about you', her former colleague said, 'is that you show up the same no matter where you are or who you're with.'

Above all, then, it's about being authentic, and about treating everyone as they would want to be treated. Jandel says that everything she knows about leadership, she learned from 25 years in the frontline in a caring profession – not least the daily dose of humility than comes from knowing you have other peoples' lives in your hands. She also uses powerful metaphors from that season of her life –

I say, 'Who is in the ICU?' 'Who particularly needs our care and attention?' You need to personalise it. What people need and want by way of care, reward and recognition is not the same for everyone. You need to understand what they value, what they like, what they need.

Amidst all this, Jandel is writing a book about all the leadership lessons she has learned as a doctor. My favourite of her proposed chapter titles is, 'Never Go To A Doctor Whose Houseplants Are Dead'.

'At the end of the day', she says, 'all of the knowledge, the plans, the processes, the rhetoric – that's all a given. As a leader, the best thing you can bring is heart.'

I couldn't agree more.

The Beginning

There is a powerful postscript to this conversation with Jandel. Around exactly the time of our first discussion, Jandel tells me later, "I sat bolt upright one morning, and realised . . . I was ready for something new." Jandel and I talked about the tapes that play in our heads sometimes, telling us as leaders what we can and can't do – which opportunities we can take, and which we can't, all of the ways in which, despite everything we've learned and role-modelled and taught, we still edit and limit ourselves. Happily, opportunity is tenacious, and it was courting Jandel. On the first of October 2018, almost five months after we first spoke, Jandel began a new role as the President and CEO of Craig Hospital, one of only two hospitals in the US dedicated to the rehabilitation of people with spinal cord and traumatic brain injuries. Jandel is the first physician to occupy the post, the first woman, and the first African-American. Jandel shared with me her first impressions after a few weeks in the role: "The hospital is a place full of hope, joy and determination. My role is still, in a way, to be a primary caregiver – I am the caregiver to the hospital, and its people." We talked about her journey to this role. She said, "I have been happy throughout my career to be a leaf in a stream. By this, I don't mean being passive; I mean being fully present, in the moment, and not constantly looking out for the next thing. I would never have got here by planning for it in advance – the right door opens once you have all the experience you need. All I ever knew was that I wanted to be a doctor – that was my guiding purpose – and everything else? Well, it just grew from there." As for me, that boy of mine, whose first months I talked about in Part One, is now thirteen years old. He is almost as tall as me, and full of wit, kindness and ambition. He is the eldest of four siblings. Some of what I see in him now, I have seen forever – his quiet watchfulness and quick smile have not changed since he was tiny. Other things are a complete surprise – his stunning artistic ability, his pretty useless spelling, his goalkeeping prowess, and the stubborn streak that runs through him like a dark coal seam. I have given my all – my energy, wit, empathy and experience – to the project of raising him thus far, and yet I feel that he is exactly who he is, regardless of the best and worst of my ham-fisted efforts. This is terrifying and thrilling in equal measure.

May the same be true for all of us in our work leading change in our organisations. May we give it our all, and then stand back and watch with delight as it happens all by itself.

And now here is my secret, a very simple secret: It is only with the heart that one can see rightly; what is essential is invisible to the eye

Antoine de Saint-Exupéry[9]

Notes

1. Buechner, Frederick, *Beyond Words: Daily Readings in the ABC's of Faith*, Harper Collins, 2004.
2. Cranfield School of Management & The Doughty Centre for Corporate Responsibility, commissioned by Coca-Cola Enterprises, *Combining Profit And Purpose: A New Dialogue On The Role Of Business In Society*, October 2014.
3. Fink, Larry, *Annual Letter to Shareholders 2018: A Sense Of Purpose*, published March 2018 and available at https://www.blackrock.com/corporate/investor-relations/larry-fink-ceo-letter
4. Brafford, Anne, *Positive Professionals: Creating High-Performing Profitable Firms Through The Science Of Engagement*, ABA Law Practice Division, 2017.
5. Fuda, Peter and Badham, Richard, 'Fire, Snowball, Mask, Movie: How Leaders Spark And Sustain Change', *Harvard Business Review*, February 2011.
6. Nadella, Satya, *Hit Refresh: The Quest To Rediscover Microsoft's Soul And Imagine A Better Future For Everyone*, William Collins, 2017.
7. Garrad, L., and Chamorro-Premuzic, T., How To Make Work More Meaningful For Your Team, *Harvard Business Review*, published on blog on 9 August 2017.
8. Snyder, Robert A., *The Social Cognitive Neuroscience of Leading Organisational Change: TiER 1 Performance Solutions' Guide for Managers and Consultants*, Routledge, 2016.
9. de Saint-Exupéry, Antoine, *Le Petit Prince*, 1943.

BIBLIOGRAPHY

Achieve Consulting Inc., *Millennial Impact Report*, June 2014.

ADP, *Achieving M&A Success*, published in 2017 at https://www.adp.co.uk/assets/vfs/Family-32/adp-files/Insights-Resources/Whitepapers/Docs/adp-unleashing-m-and-a-success-fy17.pdf, and accessed in August 2018.

Aghina, Wouter, De Smet, Aaron, Lackey, Gerald, Lurie, Michael and Murarka, Monica, 'The Five Trademarks of Agile Organizations', McKinsey, January 2018.

Ailes, Roger, *You Are The Message*, Bantam Doubleday Dell Publishing Group, 1989.

Allender, Dan B., *Leading with a Limp*, WaterBrook Press, 2006.

Amabile, Teresa and Kramer, Steven, *The Progress Principle: Using Small Wins To Ignite Joy, Engagement And Creativity At Work*, Harvard Business Review Press, 2011.

Andrews, Gail, Research paper presented in May 2015 at the Ninth Annual Conference of the Psychology Research Unit of Athens Institute for Education and Research, and summarised at www.dominican.edu/dominicannews/study-demonstrates-that-writing-goals-enhances-goal-achievement.

Applebaum, S., Malo, J., Habashy, S., Shafiq, H., 'Back To The Future: Revisiting Kotter's 1996 Change Model', *Journal of Management Development* 31(8):764–782, August 2012.

Applebaum, S., Karelis C., Le Henaff, A., McLaughlin, B., 'Resistance To Change In The Case Of Mergers And Acquisitions: Part 2', *Industrial and Commercial Training*, 49(3): 139–145, 2017.

Argyle, Michael, *Bodily Communication*, 2nd edn, Routledge, 1988.

Ariely, Dan, *Payoff: The Hidden Logic That Shapes Our Motivations*, Ted Books, Simon & Schuster, 2016.

Ariely, Dan, 2010, *What Makes Us Feel Good About Our Work*, video, TEDx, viewed April 2018.

Armenakis, A., Harris, S., Mossholder, K., 'Creating Readiness for Organisational Change', *Human Relations*, 66(6), 681–703, 1993.

Ballerini, M., Cabibbo, N., Candelier, RF., Cavagna, A., Cisbani, E Giardina, I., Lecomte, V., Orlandi, A., Parisi, G, Procaccini, A., Viale, M., and Zdravkovic, V., 'Interaction Ruling Animal Collective Behaviour Depends on Topological Rather Than Metric Distance: Evidence From A Field Study', *PNAS*, 105: 1232–37, 2008.

Baumeister, R.F., Twenge, J.M., and Nuss, C., 'Effects Of Social Exclusion On Cognitive Processes: Anticipated Aloneless Reduces Intelligent Thought', *Journal Of Personality And Social Psychology*, 2002, 83(4): 817–827.

Baumeister, Roy, Vohns, Kathleen, Aaker, Jennifer, Garbinsky, Emily, 'Some Key Differences Between A Happy Life And A Meaningful Life', *Journal of Positive Psychology*, 8(6): 505–516, 2013.

Bell, Rob, *Something to Say*, Online communications seminar programme, @ https://robbell.com/portfolio/something-to-say/ downloaded and listened to in August 2017.

Belsky, Scott, *Making Ideas Happen: Overcoming The Obstacles Between Vision And Reality*, Penguin, 2011.

Berney, Catherine, *The Enlightened Organization – Executive Tools And Techniques From The World Of Organizational Psychology*, Kogan Page, 2014.

Birkinshaw, Julian and Ridderstråle, Jonas, *Fast/Forward: Make Your Company Fit For The Future*, Stanford Business Books, 2017.

Booker, Christopher. *The Seven Basic Plots: Why we Tell Stories*, Continuum, 2005.

Blanchard, Ken and Johnston, Spencer, *The One Minute Manager*.

Brafford, Anne, *Positive Professionals: Creating High-Performing Profitable Firms Through The Science Of Engagement*, ABA Law Practice Division, 2017.

Bradley, Chris, Hirt, Martin & Smit, Sven, *Strategy Beyond the Hockey Stick – People, Probabilities And Big Moves To Beat The Odds*, McKinsey & Company, Wiley, 2018.

Bridges, William, *Managing Transitions: Making The Most Of Change*, 4th edn, Nicholas Brearly Publishing, 2017.

Bridges, William, *Transitions: Making Sense of Life's Changes*, Revised 25th Anniversary edn, De Capo Press, 2004.

Brown, Brené. *Daring Greatly – How the Courage to be Vulnerable Transforms the Way We Live, Love, Parent and Lead*, Penguin, 2012.

Bruner, Jerome, *Making Stories: Law, Literature, Life*, Harvard University Press, 2003.

Buettner, Dan, 2009, *How To Live To Be 100*, video, TEDx, viewed February 2018.

Business In The Community, *Inspiring Businesses To Improve Society Through Purpose-Driven Brands: A Purpose Toolkit*, January 2018.

Cable, Daniel, M., *Alive at Work: The Neuroscience of Helping Your People Love What They Do*, Harvard Business Review Press, 2018.

Capaldi, C., Dopko, R., and Zelenski, J., The Relationship Between Nature Connectedness And Happiness: A Meta-Analysis, *Frontiers in Psychology*, 5: 976, 2014.

Carter, Christine. *The Sweet Spot: How To Find Your Groove At Home And Work*, Ballantine Books, 2015.

Christensen et al, The Big Idea: The New M&A Playbook, *Harvard Business Review*, March 2011.

Clarke, Jane and Nicholson, Dr John, *Resilience: Bounce Back From Whatever Life Throws At You*, Crimson Publishing, 2010.

Cohen, S., Janicki-Deverts, D., and Miller, G.E., 'Psychological Stress And Disease', *JAMA*, 298(14): 1685–1687, 2007.

Coleman, John, 'You Don't Find Your Purpose – You Build It', *Harvard Business Review*, October 2017.

Columbia Threadneedle Investments, 'The UN Sustainable Development Goals: A Touchstone For Today's Responsible Investor?' *Responsible Business*, March 2018.

Coniff Allende, Sam, *Be More Pirate: Or, How To Take On The World And Win*, Penguin, 2018.

Connected Commons, 'Network Investments that Create a Sense of Purpose in Your Work', *Connected Leadership*, www.connectedcommons.com.

Cottam, Hilary, *Radical Help: How We Can Remake The Relationships Between Us And Revolutionise The Welfare State*, Virago, 2018.

Coyle, Daniel, *The Culture Code: The Secrets Of Highly Successful Groups*, Penguin 2018.

Cranfield School of Management & The Doughty Centre for Corporate Responsibility, commissioned by Coca-Cola Enterprises, *Combining Profit And Purpose: A New Dialogue On The Role Of Business In Society*, October 2014.

Crawford, E.R., Rich, B.L., Buckman, B., Bergeron, J., 'The Antecedents And Drivers Of Employee Engagement' in Truss, C., Delbridge, L., Alfes, K., Shantz, A. & Sloane, E. (eds.) *Employee Engagement In Theory And Practice*, Routledge, 2014.

Creswell, J.D., Welch, W.T., Taylor, S.E. Lucas, D.K., Gruenewald, T.L., & Mann, T. (2005) Affirmation of personal values buffers neuroendocrine and psychological stress responses. *Psychological Science* 16(11): 846–851.

Cross, R., Rebele, R., Grant, A., Collaborative Overload, *Harvard Business Review*, Jan–Feb 2016.

Cuddy, Amy, *Presence: Bringing Your Boldest Self To Your Biggest Challenges*, Orion Publishing, 2016.

David, Susan, *Emotional Agility*, Penguin Life, 2017.

Davis, Danny A. *M&S Integration – How to Do it. Planning and Delivering M&A Integration for Business Success*, Wiley, 2012.

De Bono, Edward, *Simplicity*, Penguin Life, 1999.

Dealogic, *Global Withdrawn M&A*, 01-Jan – 01 Apr, 2016.

Deci, E.L. & Ryan, R.M., 'The Importance Of Universal Psychological Needs For Understanding Motivation In The Workplace', in Gagne, M. (ed.), *Oxford Handbook Of Work Engagement, Motivation and Self-Determination Theory*, Oxford University Press, 2014.

Deloitte, *Global Human Capital Trends 2014: Engaging the 21st Century Workforce*, 2014.

Deutschman, A., *Change or Die: The Three Keys to Change at Work and in Life*, Collins, 2007.

Diener, E., and Chan, M.Y., Happy People Live Longer: Subjective Well-Being Contributes to Health And Longevity, *Applied Psychology: Health and Well-Being*, 3(1), March 2011.

Dolan, Paul, *Happiness by Design: Finding Pleasure and Purpose in Everyday Life*, Penguin, 2015.

Dunbar, Robin, *Human Evolution*, Pelican, 2014.

Dweck, Carol S., *Mindset – Changing The Way You Think To Fulfil Your Potential*, updated edn, Robinson, 2017.

Einstein, Albert, 'The World As I See It', *Forum and Century*, 84: 193–194, 1931.

Empson, Laura, *Leading Professionals: Power, Politics and Prima Donnas*, OUP, 2017 and at https://youtu.be/FaVCp58FWCQ.

Ericsson, Anders & Pool, Robert, *Peak: How All Of Us Can Achieve Extraordinary Things*, Vintage, Penguin Random House, 2017.

EY Beacon Institute, *The State Of The Debate On Purpose In Business*, 2016.

Fernandez, R., '5 Ways To Boost Your Resilience At Work' *Harvard Business Review*, June 2016.

Fink, Larry, *Annual Letter to Shareholders 2018: A Sense Of Purpose*, published March 2018 and available at https://www.blackrock.com/corporate/investor-relations/larry-fink-ceo-letter.

Frankl, Viktor, *Man's Search for Meaning*, Simon & Schuster, 1946.

Fuda, Peter and Badham, Richard, 'Fire, Snowball, Mask, Movie: How Leaders Spark And Sustain Change', *Harvard Business Review*, February 2011.

Garcia, H., and Miralles, F., *Ikigai: The Japanese Secret to a Long and Happy Life*, Penguin, 2016.

Gardner, Heidi, *Smart Collaboration: How Professionals And Their Firms Success By Breaking Down Siloes*, Harvard Business Review Press, 2017.

Garrad, L., & Chamorro-Premuzic, T., How To Make Work More Meaningful For Your Team, *Harvard Business Review*, published on blog on 9 August 2017.

Gilbert, Elizabeth, in discussion with Reid Armbruster, for the blog at www.audible.com about her book *Big Magic*, 10 August, 2016, read on 12 August 2018.

Gillespie, Nick, 'John Mackey and Conscious Capitalism Have Won the Battle of Ideas With Everyone but Libertarians', published on the Hit and Run blog on Reason.com on 15 August 2018 , accessed on 15 August 2018 – https://reason.com/blog/2018/08/15/john-mackey-and-conscious-capitalism-hav.

Godin, Seth. *Tribes – We Need you to Lead Us*, Piatkus, 2008.

Goffee, R., and Jones, G., 'Creating The Best Workplace On Earth', *Harvard Business Review*, 91: 98–106, 2013.

Grant, A., *Give And Take: Why Helping Others Drives Our Success*, W&N, 2014.

Harvard Business Review *The Business Case for Purpose*, Harvard Business Review Analytics Services, 2015.

Haslam, Alex, Social identity and the new psychology of mental health, https://www.youtube.com/watch?v=TWWZd8lrraw, watched 1 July 2018.

Hay Group, *Dangerous Liaisons*, 2007.

Heath, Chip and Dan, *Made To Stick: Why Some Ideas Take Hold And Others Comes Unstuck*, Arrow, 2008.

Heath, Chip and Dan, *Switch – How To Change Things When Changing Is Hard*, Random House Business Books, 2011.

Heffernan, Margaret, *Beyond Measure: The Big Impact Of Small Changes*, TED Books, Simon & Schuster, 2015.

Heifetz, Ronald, Grashow, Alexander, and Linsky, Marty, *The Practice Of Adaptive Leadership – Tools And Tactics For Changing Your Organization And The World*, Harvard Business Review Press, 2009.

Heskett, James, 'Putting the Service-Profit Chain to Work', *Harvard Business Review*, 2008.

Hill, P.L., and Turiano, N.A., 'Purpose in Life as a Predictor of Mortality Across Adulthood' *Psychological Science* 25(7): 1482–1486, 2014.

Hobsbawm, Julia. *Fully Connected – Surviving and Thriving in an Age of Overload*, Bloomsbury, 2017.

Jack, A.I., Dawson, A.J., Begancy K.L., Leckie, R.L., Barry, K.P., Cicca, A.H. & Snyder, A.Z., 'fMRI Reveals Reciprocal Inhibition Between Social And Physical Cognitive Domains', *Neuroimage* 66, pp385–401, 2013.

Johnston, Spencer. *Who Moved My Cheese?*

Kegan, Robert, Laskow Lahey, Lisa. *Immunity to Change: How to Overcome it and Unlock the Potential in Yourself and Your Organisation.* Harvard Business Review Press, 2009.

Kinder, Tabby, 'Olswang Three Way Merger Imminent' *The Lawyer*, 29 September 2016.

Kinder, Tabby, 'CMS to Merge with Nabarro and Olswang, Subject To Vote', *The Lawyer*, 29 September 2016.

Kinder, Tabby. 'Who Sank the Good Ship Olswang?', *The Lawyer*, 28 November 2016.

Knight, C., Haslam, S.A., Haslam, C., 'In Home Or At Home? How Collective Decision Making In A New Care Facility Enhances Social Interaction And Wellbeing Amongst Older Adults' *Ageing and Society*, 30(8): 1393–1418, 2010.

Kotter, John. *Leading Change*, HBS Press, 1996.

Krill, P.R., Johnson, R., Albert, L., 'The Prevalence Of Substance Use And Other Mental Health Concerns Among American Attorneys', *Journal of Addiction Medicine*, 10: 46–52, 2016.

Laffin, Simon, 'Sir Ken Morrison – A Retail Giant But Still Human' – blog post published on 6 February 2017 at https://laffinsdotnet.wordpress.com/2017/02/06/sir-ken-morrison-a-retail-giant-but-still-human/.

Laloux, Frederic, *Reinventing Organizations: A Guide to Creating Organizations Inspired by the Next Stage of Human Consciousness*, Nelson Parker, 2014.

Laporte, Danielle, *The Firestarter Sessions – A Soulful And Practical Guide To Creating Success On Your Own Terms*, Harmony Books, Crown Publishing Group, 2012.

Lawless, Jim, *Taming Tigers: Do Things You Never Thought You Could*, Virgin Books, 2012.

Lencioni, Patrick. *The Advantage*, Jossey-Bass, 2012.

Lock Lee, Laurence, *Smart Collaboration = Smart Money*, www.swoopanalytics.com blog, January 2017.

MacEwan, B., *Growth Is Dead: Now What? Law Firms On The Brink*, Adam Smith Esq. LLC, New York, 2013.

Mackey, John and Sisodia, Raj, *Conscious Capitalism: Liberating The Heroic Spirit Of Business*, Harvard Business Review Press, 2014.

MacLeod, D., and Clarke, N., *Engaging for Success: Enhancing Performance Through Employee Engagement*, Department for Business, Innovation and Skills, UK, 2009 – now also incorporated in a set of resources at http://engageforsuccess.org.

Marquet, L. David, *Turn the Ship Around! A True Story Of Turning Followers Into Leaders*, Penguin, 2015.

Mazutis, D., Ionescu-Somers, A., *How Authentic Is Your Corporate Purpose?*, IMD Business School, IMD Global Center For Sustainability Leadership, Burson-Marstellar, February 2015.

Meyer, Erin. *The Culture Map – Decoding How People Think, Lead and Get Things Done Across Cultures*, PublicAffairs, 2014.

Millward, David, 'Is This The End Of The Road For Traffic Lights?', *The Daily Telegraph*, 4 November 2006.

Moeller, Scott & Brady, Chris, *Intelligent M&A – Navigating the Merger and Acquisitions Minefield*, John Wiley & Sons Ltd, 2014.

Mohr, Tara, *Playing Big: A Practical Guide For Brilliant Women Like You*, Arrow, 2015.

Morrissey, Helena, *A Good Time To Be A Girl*, William Collins, 2018.

Nadella, Satya, *Hit Refresh: The Quest To Rediscover Microsoft's Soul And Imagine A Better Future For Everyone*, William Collins, 2017.

Netflix, *Netflix Culture: Freedom & Responsibility*, Steven Pappas, Operational Team Leader, Netflix, slide deck, viewed via SlideShare, February 2018.

Newport, Cal, *Deep Work*, Piatkus, 2017.

Novak, David, *Taking People With You – The Only Way To Make Things Happen*, Portfolio/Penguin, 2012.

Olswang, Simon. 'Simon Olswang: My Story', *The Lawyer*, 2 December 2016.

Patel, Ketan, *The Master Strategist: Power, Purpose and Principle*, Arrow Books, Penguin Random House UK, 2005.

Pickering, Marisue, 'Communication' in *Explorations: A Journal of Research of the University of Maine*, 3(1), 16–19, 1986.

Pink, Daniel, *Drive: The Surprising Truth About What Motivates Us*, Canongate, 2009.

Pinker, Steven, *How the Mind Works*, Penguin, 1999.

Playfoot, H., and Hall, R., *Purpose in Practice – Clarity, Authenticity and the Spectre of Purpose Wash*, Claremont Communications, 2015.

Porter, M., and Kramer, M., 'Creating Shred Value: Redefining Capitalism and the Role of the Corporation in Society', *Harvard Business Review*, 2011.

Quinn, Robert E., & Thakor, Anjan V., 'Creating a Purpose Driven Organisation: How To Get Employees To Bring Their Smarts And Energy To Work', *Harvard Business Review*, July/August, 78–85, 2018.

Quito, A., 'Netflix's CEO says there are months when he doesn't have to make a single decision', *Quartz at Work*, 19 April 2018, reporting on an interview of Reed Hastings by Chris Anderson, TED curator in Vancouver in April 2018.

Reicher, Steve & Haslam, Alex, BBC Prison Study 2006, www.bbcprisonstudy .org.

Robinson, Ken. *Out of our Minds: The Power of Being Creative*, Capstone Publishing, 2011.

Rock, David. *Your Brain at Work*, HarperBusiness, HarperCollins, 2009.

Rumelt, Richard, *Good Strategy Bad Strategy – The Difference And Why It Matters*, Profile Books, 2011.

Rutten, B.P., et al (2003) Resilience in Mental Health: Linking psychological and neurobiological perspectives. *Acta Psychiatrica Scandinavica* 128(1): 3–20.

Scarlett, Hilary, *Neuroscience for Organizational Change: An Evidence-Based Practical Guide to Managing Change*, Kogan Page, 2016.

Schawbel, Dan, 'John Mackey: Why Companies Should Embrace Conscious Capitalism', interview with John Mackey published on Forbes.com on 15 January 2013, accessed in June 2018 at https://www.forbes.com/sites/danschawbel/ 2013/01/15/john-mackey-why-companies-should-embrace-conscious- capitalism/.

Schein, Edgar H., *Organizational Culture and Leadership*, 5th edn. (The Jossey-Bass Business & Management Series), Wiley, 2016.

Seligman, M.E.P., Verkuil, P.R. & Kang, T.H., 'Why Lawyers Are Unhappy', *Cardozo Law Review*, 23: 23–55, 2001.

Senge, Scharmer, Jaworksi, Flowers. – *Presence: Exploring Profound Change in People, Organisations and Society*, Nicholas Brealey Publishing, 2005.

Simmons, Annette. *The Story Factor: Inspiration, Influence and Persuasion Through The Art Of Storytelling*, Perseus Publishing, 2001.

Sinek, Simon, *Leaders Eat Last*, Penguin, 2014.

Sinek, Simon, *Start With Why*, Penguin, 2011.

Sisodia, Raj, Sheth, Jag, Wolfe, David, *Firms of Endearment – How World Class Companies Profit From Passion and Purpose*, Pearson, 2nd edn, 2014.

Small, D.A., Loewenstein, G. & Slovic, P., 'Sympathy And Callousness: The Impact Of Deliberative Thought On Donations To Identifiable And Statistical Victims', *Organisational Behaviour And Human Decision Processes*, 103: 143–153, 2007.

Smith, Daniel. 'Why Do We Tell Stories? Hunter Gatherers Shed Light on the Evolutionary Roots of Fiction', published on The Conversation, www.theconversation.com on 5 December 2017.

Smith, Schlaepfer, Major, Dyble, Page, Thompson, Salali, Mace, Astete, Chaudhary, Ngales, Vinicuis & Migliano. 'Cooperation and the Evolution of Hunter-Gatherer Storytelling' *Nature Communications* 8: 1853, 5 December 2017.

Snyder, C.R., 'Hope Theory: Rainbows in the Mind', *Psychological Inquiry*, 2009.

Snyder, Robert A., *The Social Cognitive Neuroscience of Leading Organisational Change: TiER 1 Performance Solutions' Guide for Managers and Consultants*, Routledge, 2016.

Tett, Gillian, *The Silo Effect – Why Putting Everything In Its Place Isn't Such A Bright Idea*, Little Brown, 2015.

Tomorrow's Company Report, with Danone: *The Courage Of Their Convictions: How Purposeful Companies Can Prosper In An Uncertain World*, Mark Goyder and Norman Pickavance, 2018.

Treasure, Julian, *How To Be Heard*, Mango, 2017.

Um, Jae, 'BigLaw Partners Aren't Dumb, They're Just Not In The Room', on *Legal Evolution* blog, 24 June 2018, viewed 28 June 2018 at https://www.legalevolution.org/2018/06/big-law-partners-arent-dumb-theyre-just-not-in-the-room-054/.

Unerman, Sue and Jacob, Kathryn, *The Glass Wall*, Profile Books, 2016.

Valla, Clement, *A Sequence of Lines Traced By Five Hundred Individuals*, Vimeo Video, posted in 2011, viewed in July 2018 at https://vimeo.com/18998570.

Walker, Matthew. *Why We Sleep*, Penguin, 2017.

Webb, Caroline. *How To Have A Good Day*, Macmillan, 2016.

White, A., 'Beyond Organisational Purpose', 29 November 2016, at https://www.sbs.ox.ac.uk/school/news/beyond-organisational-purpose, accessed on 3 August 2018.

White, A., Yakis-Douglas, B., Helanummi-Cole, H., Ventresca, M., '"Saint Antony" Reflects on the Idea of Organizational Purpose, in Principle and Practice', *Journal of Management Inquiry*, 26(1): 101–107, 2016.

Whyte, David, *Crossing The Unknown Sea – Work As A Pilgrimage of Identity*, Riverhead Books, 2001.

Wood, John and McGibbon, Amalia, *Purpose Incorporated – Turning Cause Into Your Competitive Advantage*, Room to Read, 2017.

Yorke, John. *Into the Woods: How Stories Work and Why We Tell Them*, Penguin, 2014.

Zarnoth, P., & Sniedek, J.A., 'The Social Influence of Confidence in Group Decision Making', *Journal of Experimental Psychology*, 33(4), 1997.

ACKNOWLEDGEMENTS

*I have gathered a garland of other men's flowers, and nothing
is mine but the cord that binds them.*

Michel de Montaigne

My heart is overflowing with gratitude, and in particular I owe thanks to:

Annie Knight and the team at Wiley who gave me the opportunity
to write a book and then showed me how to do it. The many, many leaders and collifriends who have inspired, taught, tested and supported me
through all of the experiences that have come together to form the learning in this book – most notably at CMS. Thank you for showing me how
things work in practice, for being wholehearted, and for making work fun.

Everyone who gave their time and wisdom to this book, in particular –
Deborah Allday, Jandell Allen-Davis, Tony Angel, Sam Baker, Jason
Barnwell, Tristram Carfrae, Thomas Davies, Professor Laura Empson,
Dr Heidi Gardner, Mark Humphries, Claire Mason, Christoph Stettler
and Perry Burton, Business In The Community, and the big brains around
their table.

My alphabet soup of amazing sisters from a different mister, including
('but not limited to') – Ali Elangasinghe, Ali Tisdall, Anne Scoular, Cathy
Walton, Fran Griffith, Helen Duguid, Oonagh Harpur, Sarah David and
Shonagh Primose.

June Cormie, who gave me my first job, and Jill King, who taught me
how to sit at the table.

The DanaCup massive, for taking me to the beach. Mark, for being my truant officer and biggest fan. Nick, whose crazy idea this was and whose grammar is even better than mine.

David, whose knowledge of children's movies is now insurpassable, and above all Tommy, Frankie, Esther and Gabriel, for their almost-patience, sweet encouragement and bracing sense of perspective.

In loving memory of Wes Cameron and Juan Coto, who taught me how to climb mountains, real and metaphorical. I practise every day.

ABOUT THE AUTHOR

Jennifer Emery began her career as a corporate lawyer before moving into the business side of professional services firms in 2004. She has built her career as a coach, consultant, strategist and leader, helping organisations and individuals to unlock their potential and achieve flourishing. Jenni was Director of People at CMS from 2011 until 2016, and then Director of Strategy and Integration. Jenni joined Arup in January 2019 as their Global People Leader. She writes business papers, shopping lists and poems – only the last of these for publication. This is her first business book. Jenni lives in London with her husband and four children.

You can read more about Jenni here: https://www.linkedin.com/in/jennifer-emery-6a16685/, follow her on Twitter @jenpens, and on Instagram at @jenniemery.

INDEX